Abias Porfírio Cinco Reis

Environmental Education in Water Management

Abias Porfírio Cinco Reis

Environmental Education in Water Management

A study applied to the Cuquema River Basin, Bié-Angola

ScienciaScripts

Imprint

Cover image: www.ingimage.com

This book is a translation from the original published under ISBN 978-3-639-61934-8.

Publisher:
Sciencia Scripts
is a trademark of
Dodo Books Indian Ocean Ltd. and OmniScriptum S.R.L publishing group

120 High Road, East Finchley, London, N2 9ED, United Kingdom
Str. Armeneasca 28/1, office 1, Chisinau MD-2012, Republic of Moldova, Europe
Managing Directors: Ieva Konstantinova, Victoria Ursu
info@omniscriptum.com

Printed at: see last page
ISBN: 978-620-8-51786-1

DEDICATORY

I dedicate this work to my daughters Ariadana and Yussánide, whose unconditional love for her has been my driving force throughout this journey. To my parents, who have always believed in me and encouraged me to pursue my dreams. To my brother António Paulo Cinco-Reis, for being my constant source of inspiration and for always being by my side, even in the most challenging moments.

ACKNOWLEDGEMENTS

I would like to express my sincere thanks to everyone involved in the realisation of this scientific work. First of all, I would like to thank my dear wife Luzia Cinco-Reis for her unconditional support throughout my training, and the professors of the master's programme for the knowledge they shared.

I would also like to thank my colleagues in the academy and the department who have contributed to the progress and development of the research presented in this work.

I would also like to thank the institutions that made this project possible.

Finally, I would like to express my gratitude to my family and friends for their unconditional support and understanding during the long days of research and writing. Your love and encouragement were truly inspiring.

GENERAL CONTENTS

SUMMARY

This scientific paper addresses the intersection between environmental education and water resource management, highlighting its importance for sustainability. In the Angolan context, aspects such as water distribution and demand, agricultural and industrial use, the need for preservation and effluent treatment are discussed. Challenges such as water scarcity, drinking water supply and the impacts of climate change are explored, as well as relevant policies and legislation. The integration of environmental education into water resource management is emphasised, highlighting the role awareness raising, community involvement and behaviour change. Experiences and practices are analysed, including case studies, public participation initiatives and pedagogical approaches. Methods, tools and the importance evaluation, monitoring, innovation and sustainability in water education are explored, culminating in the need to develop educational curricula centred on sustainable water management. This work contributes to the understanding and promotion of sustainable activities in the management of water resources in the Cuquema catchment, with a focus on education as a catalyst for positive change.

Keywords: Environmental education, Sustainability, Water resources, Management, Water scarcity, Community

INTRODUCTION

The history of environmental education can be traced back to a growing concern about environmental issues that emerged in the 19th century with the Industrial Revolution. However, it was only in the 1960s that the term "environmental education" began to be widely used, in response to growing concerns about environmental degradation and the sensitisation of the population. The 1970s witnessed a significant push to establish environmental education as an institutional practice, marked by the United Nations Conference on the Human Environment in Stockholm in 1972.

Since then, Environmental Education has evolved as an interdisciplinary discipline aimed at fostering awareness, understanding and action in the face of the environmental challenges faced by society. It is also important to examine the trajectory of Environmental Education, analysing the main global events that have dealt with this issue, its institutionalisation process in Brazil and Bahia, highlighting the legal frameworks and public policies that guide its educational practice in our country and state (Holmer, 2020).

Furthermore, environmental education in the Cuquema river basin can contribute to the development of more effective and inclusive public policies. It is essential to ensure the proper management of water resources that the community actively participates in long-term sustainability. Educational programmes involving the local population can foster a deeper understanding of the interconnections between human activities and the environment, promoting a culture of environmental responsibility. Through a participatory and integrated approach, Environmental Education can help mitigate negative impacts on water resources and promote community resilience in the face of environmental and climate change.

•Presentation or brief contextualisation of the research problem.

The Cuquema River catchment area, located in Bié, Angola, plays a crucial role in supplying water to the local population and maintaining the region's aquatic ecosystems. However, this area faces significant environmental challenges stemming from disorganised land occupation and a lack of awareness about the importance of sustainable water resource management. The growing pressure on local water resources threatens water security and environmental quality, requiring the implementation of effective strategies to promote the preservation and responsible use of water.

The preservation of water resources in the Cuquema River basin is fundamental not only to guarantee continued access to drinking water, but also to protect aquatic biodiversity and related ecosystems. Degradation of these water resources can lead to negative impacts on human health, agriculture, the local economy and the ecological balance of the region. It is therefore essential to address these issues through initiatives that promote environmental awareness and community participation in sustainable water management. Considering the complexity of the challenges faced by the Cuquema River catchment area, the following starting question arises: How can we effectively promote the integration of Environmental Education into the management of water resources in this region, with the aim of increasing the level of environmental awareness among the population and guaranteeing the preservation and sustainable use of water?

•Motivation for this work:

Lack of awareness of the importance of sustainable water resource management is one of the main causes of the environmental problems faced in the Cuquema River basin. The disorderly occupation of the land for various human activities, such as agriculture, mining and urbanisation, contributes significantly to the degradation of local water resources. In , the lack of effective educational practices and public policies aimed at preserving water makes this situation even worse. It is therefore essential to promote environmental awareness and provide adequate information to the community regarding the importance of conservation and conscious water consumption.

The motivation for this work is also driven by the need to prevent and mitigate the negative impacts of water pollution on human health and aquatic ecosystems. Contamination of water resources by chemical substances, industrial waste and domestic sewage compromises water quality, making it unsuitable for human consumption and damaging the biodiversity of rivers and lakes. Therefore, investigating and understanding the main types of pollution and their sources in the Cuquema River basin is essential for developing effective strategies for managing and protecting water resources.

In addition, the research is motivated by the opportunity to contribute to the promotion of environmental sustainability and the socio-economic development of the region. Integrating environmental education into water resource management can not only sensitise the population to the importance of water conservation, but also encourage the adoption of sustainable practices and the active participation of the community in decision-making regarding the use of natural resources. In this way, it is hoped that this study can provide extremely important data for the development of public policies, educational programmes and community actions aimed at the protection and sustainable management of the Cuquema River basin.

Justification:

This research is important because the Cuquema river basin, the first tributary of Angola's main river (the Kwanza River), is extremely important for supplying water to the Angolan population. The region is a totally plateaued river in accordance with its altitude, receiving large volumes of annual rainfall and without episodes of drought. However, today it has environmental problems as a result of intensive occupation of the land for various activities and a lack of awareness and knowledge of the importance of sustainable management of water resources in the basin under study, as well as the implementation of effective educational methods that promote the preservation and responsible use of this resource. Thus, environmental education actions can enable interventions and promote community participation, seeking to achieve a more rational and less degrading use of the region's water resources.

After several searches in the study area and with institutions such as the Angolan National Institute of Water Resources (INRH), created in 2010 by Presidential Decree No. 253/10 of 16 November, whose mission is to ensure the implementation of the guidelines of the national water resources policy, as well as the municipal administration of Cuito-Bié, there has never been a solution from a practical point of view or scientific research related to the proposed topic.

This study aims to raise awareness of the importance of environmental education for the preservation, sustainable use and conservation water in the Cuquema river basin. It also aims to analyse the new paradigm of Integrated Water Resources Management (IWRM), in accordance with the legislation in force in Angola, and the role of river basin committees. The research also involves the participation of the local community, contributing to the environmental sustainability of the region.

•General objective

To assess the level of environmental awareness of the population of Cuquema in relation to the management of water resources.

- **Specific objectives**

a) Analysing the practices preservation e conservation of water resources in the town of Cuquema.

b) Identify o level of knowledge of the population related the impacts pollution on water quality.

c) Diagnose the main types of human activities that have an impact on water quality.

Work structure.

The work is structured in eight chapters that deal with various aspects of integrating environmental education into water resource management. The introduction outlines the challenges and objectives of the research.

Chapter 1 explores the theoretical foundations of environmental education. Chapter 2 focuses on water resource management, looking at principles, policies and the challenges faced.

Chapter 3 analyses the integration of environmental education into water resources management, based on practical experiences. Chapter 4 presents various environmental education and water resource management practices. Chapter 5 describes the methodology adopted in the research.

The results are presented in Chapter 6, followed by the discussion in Chapter 7. Finally, Chapter 8 presents the general conclusions of the study, highlighting its contributions and recommendations for future research.

THEORETICAL FRAMEWORK

1. FUNDAMENTALS OF ENVIRONMENTAL EDUCATION

Environmental education is a broad and multifaceted discipline that encompasses several aspects crucial to its understanding and effective implementation. Initially, it is essential to explore the brief history of this area in order to understand its origins and evolution over time. Next, it is essential to examine the principles and concepts that underpin this educational approach, as they provide the theoretical basis needed to guide practices and interventions. In addition, the author uses foundations from the fields of critical education, Environmental Education and Information and Communication Technologies to theorise and problematise the relationship between Environmental Education and Virtual Environments, from the perspective of the dynamics of cyberspace (Freitas, 2023).

Furthermore, understanding the connection between environmental education and sustainability is crucial to recognising the role of this discipline in promoting more sustainable behaviours and lifestyles. By considering Environmental Education in the Angolan context, it is possible to analyse how this educational approach adapts and responds to the specific environmental challenges faced by the country, offering important information for the practical implementation and development of effective environmental policies. It is important to emphasise that, in this context, human interactions in virtual environments are not without intentionality, influencing processes and dynamics that shape reality (Freitas, 2023).

1.1. A brief history of environmental education

Santos (2024) emphasises the crucial importance of environmental education in guiding the learning process and promoting reflection on society's responsibility for development. sustainable. Throughout history, the need for specific legislation, such as Environmental Education, has become evident due to industrialisation and the resulting environmental impacts. The International Conferences on Environment and Development have played a significant role in raising awareness and addressing global environmental issues, especially in the face of the technological advances of the Third Industrial Revolution.

The author emphasises the importance of social transformation guided by Environmental Education, which should be integrated in an interdisciplinary or transdisciplinary way at all educational levels, covering all age groups. He argues that the effective implementation of Environmental Education in the school curriculum is essential to achieve transformative actions that meet the needs of all social strata. In addition, this expressiveness was greater than that identified in newspapers. On the social network, there were more results related to "accounts", i.e. profiles on the network; while the search for "tags" resulted in a low expressiveness of Environmental Education as a term mentioned on the social network (Freitas, 2023).

Environmental education is a field dedicated to promoting awareness, knowledge and action in favour of the preservation and sustainability of the environment. Over time, it has gone through several important phases and milestones that have influenced its evolution and relevance in society. This text will look at the main events and documents that have contributed to its development, emphasising the importance of avoiding misconceptions and practices that are distant from the theoretical, scientific and literary bases pertinent to the subject (Freitas, 2023).

The 1960s and 1970s saw the emergence consolidation of environmental education. In this context, significant social and cultural movements influenced environmental awareness. Of particular note was the impact of Rachel Carson's book "Silent Spring", which warned of the damage caused by human activity, especially the indiscriminate use of chemical products (Santos, 2024).

In 1968, UNESCO defined the environment as an interaction of social, cultural and economic aspects, initiating environmental studies focussed on the physical environment. The year 1972 was marked by the United Nations Conference on the Human Environment in Stockholm, an important milestone that put the environmental issue on the global agenda. Then, in 1975, the Belgrade Charter resulting from the Intergovernmental Conference on Environmental Education in Tbilisi established fundamental principles and strategies for environmental education, addressing physical, chemical and biological aspects and the transformations caused by human actions in social dynamics and cultural processes. In the context of cyberspace, Environmental Education assumes a relevant role as an object of promotion and dissemination. When present in cyberculture, it enables the promotion and problematisation of sustainability, as well as contributing to an understanding of the Virtual Environment and its unique complexity (Freitas, 2023).

In the 1970s, concern about sustainability and the survival of the planet grew, highlighted by the publication of "A Blueprint for Survival", which proposed measures to promote a healthy and balanced environment, warning of the limits of the Earth's finite resources. Furthermore, the Virtual Environment, which reflects the dynamics of everyday human life and mediates social relationships and processes, plays a crucial role in the construction of reality. It is therefore important to consider that the advance of digital technological resources is intrinsically linked to human and social development in new spaces, virtual spaces (Freitas, 2023).

In the years since, environmental education has expanded and deepened. During the 1990s, important international events took place, such as the Second United Nations Conference on Environment and Development, which resulted in the signing of the Treaty on Sustainable Societies.

Sustainable Development and Global Responsibility, as well as the enactment of national environmental education policies in various countries. This decade also saw the recognition of the environment as a subsidy for environmental management, aimed at appropriate planning for the preservation and conservation of biodiversity, including geography and socio-diversity, for the construction of a plural and equitable geographical space (Guerra, 2022).

In 1992, Rio-92 reinforced the importance of this field as a tool for sustainability. In 1997, UNESCO's Thessaloniki Declaration emphasised the need for an integrated approach to environmental education in society. The spotlight on Environmental Education continued in 2005, with its presence in various spheres, through the publication of contemporary studies and practices aimed at promoting the sustainability of social and productive practices. This period was marked by the quest to suppress economic rationality with the environmental paradigm, promoting a new relationship between society and nature (Silva, 2020).

In the 2000s, the National Curriculum Parameters included the environmental dimension as a cross-cutting theme in school curricula, representing a significant advance in the integration of Environmental Education into formal education. In this context, Environmental Education has become a dynamic and participatory educational mechanism, in which educators and students take the lead in the teaching and learning relationship, building the knowledge necessary for perception, diagnosis and sustainable socio-environmental interventions (Silva, 2020).

In the 2010s, awareness of climate change fuelled advances in Environmental Education around the world, with the implementation of public policies and institutional partnerships aimed at sustainability. This reflects an effort aimed at consolidating and internalising a new paradigm, the environmental paradigm, through Applied Environmental Education (Guerra, 2022).

In the 2020s, Environmental Education gained even more prominence globally, with climate change and the COVID-19 pandemic highlighting the urgent need for an integrated approach between human health and environmental preservation. Online education, driven by the pandemic, has brought about changes in educational systems, highlighting the interconnection between human health and the health of the planet. Environmental Education acts as an instrument for changing mentality and behaviour, through dialogue and active methodologies, helping in the design, elaboration and participatory implementation of management, recovery, conservation and environmental preservation plans (Freitas, 2023).

Today, Environmental Education continues to play a crucial role in raising awareness and engaging society in environmental issues arising from anthropogenic factors, such as the different types of use of nature and the gynaecological impacts of human activities. However, it faces challenges such as the need for multidisciplinary approaches, the integration of different audiences and the promotion of sustainable practices. At the same time, growing concerns about the environment and the search for sustainable solutions offer opportunities for the expansion and deepening of Environmental Education (Guerra, 2022).

Environmental Education has evolved over time, going through different phases and milestones that have contributed to its consolidation. In relation to Virtual Environments, it is important to observe the implications of the production and use of digital technologies on interaction, dialogue, (in)formation and communication processes. Recognised as an essential tool for promoting sustainability and preserving the environment, Environmental Education is considered a "philosophy of life". Through educational initiatives and sustainable practices, Environmental

Education continues to play a fundamental role in building a society that is more aware of and engaged with environmental issues (Freitas, 2023).The Principles and Concepts that guide Environmental Education are fundamental to its effective practice and application. These include interdisciplinarity, which recognises the complexity of environmental issues and the need for integrated approaches, and participation, which emphasises the active involvement of communities in decision-making and the implementation of environmentally sustainable actions. In addition, concepts such as ecology, which emphasises the interactions between organisms and their environment, and ethics, which promotes values of responsibility and care for the environment, are fundamental to a comprehensive understanding of Environmental Education. The environmental education process must be approached holistically, recognising its dynamism and constant construction, with the aim of strengthening the political and socio-environmental dimension of Environmental Education (Neiman, 2022).

Angolan laws on environmental education and the environment play a fundamental role in protecting and preserving the environment, as well as in raising citizens' awareness and action in relation to environmental problems. According to Costa, (2022), some important legislation and guidelines include:

-The Basic Environmental Law, which establishes the principles of environmental protection and preservation, as well as the rational use of natural resources in Angola.

- The Environmental Impact Assessment Law, which regulates the assessment of the environmental impacts of human projects and activities, with the aim of minimising negative effects on the environment.

- The Water Law, which deals with the management and protection of water in Angola, including the conservation of rivers, lakes and other bodies of water.

- The National Environmental Quality Programme, which aims to implement measures in accordance with the principles of environmental protection and preservation.

- The Basic Law of the Education System, which defines the principles of education in Angola, including the integral formation of the individual and the construction of a free, democratic and progressive society.

In addition, Angola recognises its importance and environmental responsibility at international level, having signed and ratified various international agreements to preserve environmental quality.

In this way, Angolan laws on environmental education and the environment establish principles and guidelines for environmental management and education, with the aim of minimising negative effects on the environment and promoting sustainability.

According to Lubanzadio (2020), the principles and concepts of environmental education include:

- Dynamic, permanent and participatory training, seeking to transform people into active agents responsible for finding solutions to environmental problems.

- Raising awareness, sensitising society to environmental problems and possible solutions.

- Interdisciplinarity, involving different disciplines and perspectives in the study and

approach of environmental problems.
- Sustainability, promoting the conservation and sustainable use of natural resources and the environment.

The objectives of environmental education include environmental conservation, recognising the environment as a good for the common use of the people and the search for quality of life and sustainability.
Finally, Environmental Education involves not only theory, but also concrete practices and actions for the conservation and improvement of the environment, such as ecotourism, community action and non-formal education.

1.2. The relationship between environmental education and sustainability

The relationship between Environmental Education and Sustainability is essential for tackling contemporary environmental challenges. Environmental Education plays a key role in promoting sustainability, empowering individuals and communities with the knowledge, skills and attitudes needed to adopt more sustainable practices and behaviours. In turn, sustainability provides a context and purpose for Environmental Education, seeking to balance the present and future needs of people and the planet. Collaboration between these two fields is crucial to promoting a more resilient and equitable future. However, there is a lack of work on campus with these themes, indicating that students are not experiencing these discussions during their initial training. This lack may discourage students from researching the subject (Silva A. F., 2021). (See fig. 1.1).

Figure 1.1- Environmental education influences sustainable development. Source: Author.

Environmental education influences sustainable development by addressing environmental problems and interacting with the political, social, economic and environmental spheres. By integrating these spheres, it contributes to a more complete sustainable development, making the education system more relevant and realistic (Bastos, 2021). Currently, global economic growth is one of the challenges in building sustainable development that adequately values human and natural resources, promoting their perpetuity (Pinho, 2021).In basic education and universities, environmental content is covered in textbooks and scientific articles, with the reinforcement of laboratory research and other methodologies that contextualise topics related to environmental study. In addition, environmental education can influence teacher training, with a view to building integrative knowledge and promoting a dialogue that allows individuals to be autonomous in their institutions (Cachapa, 2020).

When analysing the relationship between environmental education and sustainability, it is necessary to consider the challenges and perspectives that emerge. Some of these he challenges include the need for a growing internalisation of environmental issues, knowledge that is still under construction, and the importance of strengthening integrative visions that encourage reflection on diversity and the construction of meanings around individual-nature relations, global and local environmental risks and relations between the political, social, economic and environmental spheres. Costa, K. B. (2022). he relationship between environmental education and sustainability is fundamental to citizens' awareness and action in relation to environmental problems and the promotion of sustainable solutions. Environmental education can influence sustainable development through its targeted approach problem-solving and the interaction between the political, social, economic and environmental spheres Cachapa (2020). However, it is necessary to consider the challenges and perspectives that emerge in the search for sustainable solutions, such as the need for growing internalisation of environmental issues and the importance of strengthening integrative visions that encourage reflection on diversity and the construction of meanings around individual-nature relations, global and local environmental risks and relations between the political, social, economic and environmental spheres.

1.3. Environmental education in the Angolan context

Environmental education in the Angolan context faces unique challenges and opportunities due to the diversity of ecosystems and the country's specific needs. Angola has a rich biodiversity and natural resources, but also faces significant threats such as deforestation, pollution and soil degradation. In this scenario, Environmental Education plays a crucial role in sensitising the population to the importance of environmental conservation and promoting sustainable practices. At the same time, it is important to consider Angola's cultural and socio-economic specificities when developing Environmental Education strategies and programmes, ensuring that are contextually relevant and effective in addressing the environmental challenges facing country. Nadina, L. H. (2020). (See fig. 1.2).

Figure 1.2- Environmental education influences sustainable development. Source: Author.

A study on environmental education in Angola by Ruth Gabriel Canga Buza investigated the ideas, knowledge and practices reported by teachers in a country under reconstruction. The study revealed that environmental education is seen as an educational concern, but has not yet been implemented as public policy and continuing teacher training. In addition, the research identified the need for a cultural change so that environmental education can be constituted as a cultural process from the context of origin to the national context, another panorama needs to be woven, to verify the acceptance of international guidelines by national legislation, bodies and environmental educators (Buza 2013).

Another study, conducted by Manuel Kamuenho Alberto, looked at environmental education in the pedagogical process in Angola. The author argues that environmental education should be seen as the right of citizens to educate themselves in order to protect themselves from the risks resulting from the irrational exploitation of natural resources. In addition, he emphasises the importance of environmental education in the pedagogical process at Cuito Cuanavale University in Angola. the aim of training students to participate responsibly and competently in society (Alberto, 2018).

Environmental education is also addressed in the context of sustainable development and environmental preservation in Angola. The article emphasises the importance of environmental education in the school context, with a focus on national environmental policy and social service. It emphasises the need to improve environmental education in order to guarantee access to and permanence in school and to raise the cultural level of the population in order to understand scientific and technological advances (Silva P. S., 2021).

In Angola, environmental education is a right guaranteed by the Constitution of the Republic, in Article 39, which states that everyone has the right to live in a healthy and unpolluted environment, as well as the duty to defend and preserve it. However, the current Angolan education system has several weaknesses, such as the low level of parental participation in the education process, the outdated physical, methodological and didactic structure in schools, and the lack of investment in the sector. In addition, most of the environmental problems experienced today are directly related to modern lifestyles, characterised by excessive consumerism and

unsustainable practices (Freitas, 2023).
Environmental education in Angola is crucial to developing environmental awareness and promoting sustainability. However, the country faces challenges in implementing public policies and training teachers. Cultural change is essential to overcome these challenges and improve environmental education in schools. In addition, many of today's environmental problems are linked to modern lifestyles, characterised by excessive consumerism and unsustainable practices. This situation highlights the importance of prioritising environmental education at all levels of education and in the non-formal sphere, despite the financial limitations faced by government bodies (Silva, 2021).

In the Angolan context, Environmental Education faces unique challenges and opportunities due to the diversity of ecosystems and the country's specific needs. While Angola has a rich biodiversity, it also faces threats such as deforestation, pollution and soil degradation. In this scenario, Environmental Education plays a crucial role in sensitising the population to the importance of environmental conservation and promoting sustainable practices (Freitas, 2023).
However, it is essential to consider Angola's cultural and socio-economic specificities when developing Environmental Education strategies and programmes, ensuring that they are contextually relevant and effective in addressing the environmental challenges faced by the country. By empowering individuals and communities to play an active role in the preservation and sustainable use of natural resources, we can contribute to a healthier and more prosperous future for all Cachapa (2020).

CHAPTER II
WATER RESOURCES MANAGEMENT

The effective management of water resources is fundamental to environmental, social and economic sustainability. Chapter II of this document focuses Water Resources Management, addressing crucial aspects related to the distribution, demand, preservation and sustainable use of water. Various topics will be analysed throughout this chapter, from the different uses of water to the fortuitous challenges faced in managing this vital resource.

Fully understanding these topics is essential to guide policies, practices and interventions aimed at guaranteeing the availability and quality of water for present and future generations. In the context of water resource management, the difference in gender roles influences the way water is managed and accessed. The existence of unequal power and resource relations means that men are more able to access and claim their rights to water resources (Rosa, 2020).

2.1. Distribution and demand

The Distribution and Demand of water are fundamental aspects to be considered in the management of water resources. Costa, K. B. (2022). Cachapa, A. F. (2020). Distribution refers to the availability of and access to water in different regions and communities, while demand relates to the consumption of water for various purposes. Understanding the geographical distribution water and the varying demands helps to target effective management and allocation strategies for this vital resource (see fig.1.3).

Figure 1.3- Water distribution project. Source: Author.

In the context of water distribution, several factors contribute to inequality in supply, such as discontinuity of supply, operational and design problems in the distribution network, lack of sanitary sewerage and inadequate maintenance of the infrastructure. These challenges further exacerbate water losses throughout the distribution system. (Lubanzadio, 2020) Summarising the information from the sources on Water

Resources, we can highlight:
1. Importance of Water Resources: Water is essential for human life and for most human activities. Angola has an abundance of water resources, but access to drinking water is still a privilege for the few.
2. Unequal distribution: there is an unequal distribution of water resources on the planet, with regions with high availability and others with scarcity. Fresh water, which is fundamental for economic development and quality of life, represents only a small part of the total water available.
3. Types of Water Resources: water resources include surface water, groundwater, rainwater, coastal water and frozen water, each of which plays a specific role in environmental sustainability and human use.
4. Water Resource Management: The effective management of water resources is crucial to ensure their sustainable use, including conservation policies, water treatment, pollution control and strategies to deal with extreme events such as droughts and floods (Lubanzadio, 2020).

The unequal distribution of water and the growing demand for drinking water represent significant challenges for environmental sustainability and human development. Effective management of water resources, taking these challenges into account, is crucial to ensuring equitable access to water, conservation of aquatic ecosystems and long-term environmental sustainability (Silva P. S., 2021).

Figure 2.4 - Domestic water use.

Source: Own authorship.

2.1.1. Domestic water use

Domestic water use encompasses a variety of everyday activities, such as personal hygiene, house cleaning and food preparation. This category represents a significant portion of total water demand and it is therefore crucial to implement efficient water use and conservation practices in homes and communities to ensure the sustainability of this resource.However, despite the importance of domestic water use, their participation in discussions about water use is still limited (Guarda, 2019). (See fig. 2.4).

In Brazil, for example, domestic consumption is the third largest category of water use, accounting for 8 per cent of the total. In contrast, in underdeveloped countries, domestic use represents only 8 per cent of the total, while agriculture holds the largest share, with 82 per cent of consumption. Regardless of the location, households are the places most affected by water shortages (Viera, 2010).

It is important to emphasise that water and sanitation projects can lead to significant improvements in access to citizenship and in the living conditions of communities. For Alves (2020), in areas where water is scarce, such as the Mosquitia region in Honduras, water management and treatment at home are fundamental to guaranteeing safe drinking water. Although piped water reaches homes, it is not always safe for consumption. Everyone in the communities has some knowledge of the importance of water management and treatment, but this knowledge is not always applied.

To ensure the safety of drinking water in these areas, it is crucial to promote awareness of the importance of treating water at home. This can be done through training meetings, radio programmes, posters and other educational materials. It is essential to store treated water in clean and safe containers, such as plastic barrels or water tanks (Nadina, 2020). , in regions where water is abundant, it is important to practise conscious water consumption, avoiding waste and reducing consumption wherever possible. This includes simple attitudes such as turning off the tap when brushing your teeth, taking short showers and avoiding high-pressure showers, among others. The health authorities also warn against using treated water.

Domestic water use is common and involves a range of daily activities. In areas with water scarcity, it is crucial to promote the proper management and treatment of water to ensure its safety for drinking. In regions where water is abundant, it is important to practise conscious consumption to avoid waste (Nadina, 2020).

Domestic water use in Angola is a relevant issue, especially given the scarcity of water in some regions of the country. Although the availability of drinking water for the urban population is 72%, the majority of inhabitants in Luanda, the capital, depend on cistern trucks for their water supply, with only part of the population having access to piped water. Furthermore, the country still only exploits around 5% of its water potential for socio-economic development (Lubanzadio, 2020).

The research carried out in the Mosquitia region of Honduras could be relevant for Angola, especially considering the importance of treating water at home to make it safe to drink. The research revealed that although the ethnic groups studied have some knowledge about the importance of water treatment, this knowledge is not always applied. In addition, most families carry out some form of water treatment at home, but there are variations in the application of this knowledge between the different ethnic groups (Costa, 2022).

Therefore, promoting water treatment at home can be an important measure to ensure the safety of drinking water in Angola, especially in areas where access to drinking water is limited. In addition, it is essential to consider the participation of all social groups in the formulation of water management policies and strategies, ensuring an inclusive and equitable approach. (Silva A. F., 2021)

In Angola, water treatment at home can be carried out using different methods, such as boiling, exposure to the sun or the addition of chlorine. However, the cost of chlorine can be a barrier for some people, especially if it is not provided free of charge by health services. Treated water is usually stored in containers such as barrels at home (Sassoma, 2013).

Lack of awareness about the quality of piped water can also be a problem in Angola, especially in areas where the population relies on cistern lorries for their water supply. It is therefore crucial to promote awareness of the importance of water treatment at home and to provide teaching materials and training programmes to guarantee the good health and safety of drinking water. Water quality classification and control studies have been carried out in Angola, as demonstrated in Sassoma's 2013 dissertation, which characterised the water of the Catumbela River. These studies are fundamental to ensuring that the water consumed by the population meets safety and potability standards (Sassoma, 2013).

As can be seen, domestic water use in Angola is a relevant issue, especially considering the scarcity in some regions. It is essential to promote water treatment at home, raise awareness about water quality and guarantee access to safe drinking water through public health programmes and education initiatives (Nadina, 2020).

2.1.2. Agricultural water use

Agricultural water use plays a fundamental role in food production and the livelihoods of farming communities. However, this use consumes a significant amount of water, mainly for irrigating crops and raising livestock. It is therefore crucial to implement sustainable agricultural management strategies to optimise the use of this finite resource (Nadina, 2020). An important aspect of sustainable agricultural management is the use of efficient irrigation techniques, which help to maximise the use of available water. In addition, choosing crops that are better adapted to the local climate can reduce the need for irrigation water, contributing to the conservation of this resource (Costa, 2022). (See fig. 2.5).

Figure 2.5 - Agricultural water use.

Source: Author.

However, it is important to emphasise that the agricultural use of water can have environmental impacts, such as increasing the concentration of metals in agricultural soils. Studies, such as the one carried out by Frascareli in 2021, have identified alarming concentrations of metals in areas of agricultural use, exceeding the guidelines set by environmental authorities. This emphasises the importance of sustainable agricultural practices and environmental monitoring to mitigate these negative impacts.

The agricultural use of water is essential for food productionbut it requires careful and sustainable management to guarantee the long-term availability of this resource and minimise environmental impacts (Frascarelli, 2021).

The agricultural use of water plays a significant role in Angola, especially in the south of the country, where it is essential for irrigating pastures. Groundwater is the main source of supply for the irrigation, with most of Angola's known water systems relying on it. However, groundwater utilisation varies considerably across the country, with some provinces using a much higher proportion of this resource than others (Costa, 2022).

In the reservoir's catchment area, the main soil classes are dystrophic red-yellow clay soils and cambisols, as identified by Frascareli in his 2021 study. In addition, water is essential for agricultural production in Angola, representing a significant portion of the country's total water use. Angola's National Water Plan is focused on improving the efficiency of water use in agriculture, which includes initiatives such as the construction of dams to regularise water flow and improve irrigation (Nadina, 2020).

In addition, wastewater is also utilised in agriculture as a way of mitigating problems related to water availability and basic sanitation. This can include plant production projects, irrigation or drainage systems, as well as water and soil conservation and management practices, as mentioned by (Freitas, 2021).

The inappropriate use of water, especially in rural areas, can have significant impacts on the quality and quantity of natural water resources in Angola, potentially leading to a shortage of drinking water for the population. On the other hand, water reuse is a practice used in the country to irrigation needs and improve basic sanitation. Lubanzadio, H. N. (2020). Water reuse is also an important practice for the supply of drinking water, especially in regions where the availability of water resources is limited. However, it is crucial to ensure that this reuse is carried out safely and efficiently in order to avoid contaminating water resources. The agricultural use of water is one of the main uses of this resource in Angola, especially in the south of the country, where groundwater is widely used for irrigation. Angola's National Water Plan aims to improve the efficiency of this use, including the construction of dams to regularise the flow of water and improve irrigation practices. However, it is important to be aware of potential negative impacts, such as soil contamination due uncontrolled agricultural practice, as highlighted by Frascareli in his study (Frascareli, 2021).

Water reuse is, in fact, an important practice in Angola to irrigation and sanitation needs, especially in areas where access to drinking water is limited. The use of water trucks to supply water is a common practice in many regions of the country, including Luanda, the capital (Nadina, 2020).

However, it is important to emphasise that the inappropriate use of water, including reuse practices without adequate control or management, can have negative consequences for the quality and quantity of natural water resources. It is therefore essential that the authorities and the population adopt measures to promote the sustainable use of water and guarantee the preservation of this vital resource for future generations (Lubanzadio, 2020).

2.1.3. Industrial use of water

Industrial water use covers a variety of industrial processes, such as cooling, energy production and product manufacturing. Industries can be major consumers of water and generators of contaminated effluents, posing significant challenges for the management of water resources. (Rodríguez-Guerra, 2020) The implementation of water treatment technologies and the adoption of social and environmental practices are essential for the sustainable management of water resources. cleaner production methods are essential for reducing the environmental impact of industrial activities and promoting the sustainable use of water.

The industrial use of water is a crucial aspect of sustainable industrial production, with reuse and recycling being essential tools for water management in industries. Water use in the construction industry is also significant, and ensuring an adequate supply of water and its treatment is fundamental to the sustainability of the sector.

Treatment and reuse of water in industrial processes can bring environmental and economic benefits, including reduced water consumption and lower costs for acquiring and treating . Water reuse can be direct or indirect, indirect reuse occurring through the use of treated wastewater for irrigation, which can then seep into the water table and be used by industries (Nadina, 2020)

The industrial use of water is subject to regulations and licensing, which aim to guarantee environmental protection and pollution prevention. This licensing process generally includes an assessment of the environmental impact of industrial activity, the implementation of measures to reduce pollution and the monitoring of compliance with licence conditions (Freitas, 2023).

Regulations vary according to the type of industrial activity and the potential environmental impact, with some activities requiring a more detailed assessment and stricter conditions. The aim is to ensure that industries adopt sustainable practices and minimise their impact on the environment (Costa, 2022).

In addition to government regulations, there are initiatives to promote sustainable water management in industries. For example, in Mozambique, the he National Water Resources Management Strategy aims to guarantee the protection and sustainable use of water resources throughout the country. This strategy includes measures to promote the efficient use of water, reduce water losses and protect water quality (Cachapa, 2020).

It is important to emphasise that the participation of local communities and private sector actors is fundamental to the success of these initiatives. Collaboration between different stakeholders is essential to ensure the effective and sustainable

management of water resources, both in the industrial context and at national level (Cachapa, 2020).
The industrial use of water in Angola is a relevant issue, especially considering the context of water scarcity and the importance of economic development. Several articles emphasise the importance water for socio-economic development and the need for sustainable management of water resources (Freitas, 2023).
Angola's Water Law establishes the general principles of the legal regime relating to the use of water resources. This legislation applies to both surface water and groundwater, which form an essential part of the national hydrological cycle. Access to water is regulated through common and private uses, and rainwater drainage and liquid waste sanitation are also subject to specific regulations Frascareli, D. (2021).
Water is fundamental in various sectors in Angola, such as drinking water supply, wastewater sanitation, agriculture, industry, hydroelectric power generation, among others. This variety of uses emphasises the importance of proper and sustainable management water resources, with the aim of guaranteeing the availability and quality of water for the present and future needs of the Angolan population and economy (Nadina, 2020).

Water management in Angola faces a number of environmental challenges that directly impact the availability and quality of water resources. These challenges include the excessive use of pastures, which results in soil erosion and contributes to desertification, deforestation of the rainforest, soil erosion that pollutes the waters and leads to the silting of rivers and dams, as well as inadequate drinking water supplies (Monteiro, 2022).
Like the Federal District, other regions also have specific bodies responsible for water management. These bodies are tasked with planning and implementing actions aimed at preserving the quantity and quality of water. However, the effectiveness of these measures is often hampered by a lack of resources, adequate infrastructure and coordination between the different sectors involved in water management (Monteiro, 2022).
Faced with these challenges, it is crucial that more efficient and sustainable water management policies and practices are implemented, which take into account not only immediate needs, but also the long-term conservation of water resources and environmental balance. This requires a joint effort by the government, civil society and the private sector, aimed at guaranteeing equitable access to water and the protection of aquatic ecosystems for present and future generations (Monteiro, 2022).
Water scarcity represents a significant challenge in the southern areas and coastal regions of Angola, where extreme droughts in recent years have affected the availability of water for both industrial use and the general population. The Pure Aqua company offers affordable and effective solutions for water treatment in these regions, employing a variety of methods such as ultrafiltration systems, water media filters and reverse osmosis for brackish water. The issue of the industrial use of water in Angola is multifaceted, involving legal, environmental and economic aspects (Costa, 2022). The sustainable management of water resources is vital for the country's socio-economic progress, especially considering water scarcity and the growing demand for quality water by both industry and the population. Water scarcity

is a crucial concern due to the lack of investment, which negatively affects daily life and maintenance capacity, as exemplified by the challenge faced by a school with hundreds of students and staff. Dealing with the industrial use of water requires a holistic approach that encompasses sustainable management practices, appropriate regulations and the promotion of innovation and technology transfer. In the construction sector in particular, ensuring an adequate supply of water and its sustainable management are essential for sustainability, emphasising the importance of water reuse and recycling. As mentioned by Monteiro (2022), water is life and hygiene, its scarcity can trigger health problems and even affect access to food.

2.2. The need for the preservation and sustainable use of water

The importance of conserving and using water sustainably is a global concern due to its limitation and the adverse impacts resulting from its excessive use. Ensuring the availability of this resource for generations to come requires the promotion of its preservation and the adoption of sustainable practices in all human activities. This includes not only the conservation of water bodies, but also the responsible and efficient use of this resource in all our activities. As emphasised by Silva (2022), education about the need to preserve water is not a uniform approach, but a need adapted to the particularities of each context and community (see fig. 2.6).

Figure 2.6 - Poor sustainable water conservation. Source: Own authorship.

The preservation and sustainable use of water is essential to ensure that this vital resource is available for present and future generations. It is crucial to adopt practices that promote water conservation, from its collection and treatment to the treatment of wastewater before its return to water bodies, in order to ensure its availability and quality in the long term. As mentioned by Silva (2022), the UN has emphasised the importance of sustainable access to water and sanitation as a fundamental human right, putting pressure on member countries to implement public policies that guarantee universal access. In agro-industry, where water plays a crucial role, preservation becomes even more critical, given the small percentage of water suitable for consumption. It is important not only to provide water and sewage infrastructure, as Silva (2022) emphasises, but also to guarantee effective access to these services, without economic restrictions based on the consumer's social status. To promote the sustainable use of water, simple measures such as preventing the pollution of rivers, lakes and beaches, promoting conscious consumption and educating the community about environmental issues are essential. Such actions can contribute significantly to improving the current situation and guaranteeing a more promising future for all, as Silva (2022) emphasises.

In Angola, the preservation and sustainable use of water. This is even more important given the difficulties faced due to water scarcity and the challenges in managing water resources. Despite the country's water potential, access to drinking water is still precarious, below the standards recommended by the World Health Organisation. Issues related to legal certainty and the need for investment in the sector are also highlighted as significant challenges by Silva (2022).

The quality and cost of water are critical factors in Angola, with many deficiencies in quality and high prices, mainly affecting the most vulnerable sections of the population, who often have to resort to the informal water market. In addition, basic sanitation coverage is limited, which has direct impacts on the health and development of the population, resulting in high infant mortality rates. Water, being an essential element for the survival of all living beings, is intrinsically linked to the social right to health. However, not only the social right to health, but also basic sanitation, plays a crucial role in promoting well-being, as pointed out by Silva (2022).

The severe drought experienced in some Angolan provinces, such as Cunene and Namibe, highlights the urgency of adopting measures to guarantee sustainable water management and promote the rational consumption of this vital resource. The population in these regions faces daily difficulties in accessing water, emphasising the importance of preserving and conserving water to ensure the continuity of life and the well-being of present and future communities. As mentioned by Monteiro (2022), guaranteeing drinking water and sanitation for all is essential for sustainable water management and the well-being of populations.

In this scenario, it is crucial that Angola strengthens measures to guarantee the prudent use of water, promoting efficient and rational consumption practices, especially in the face of population growth and pressure on water production and consumption levels. Awareness and concrete actions in favour of the preservation and sustainable use of water are fundamental to guaranteeing a safer and healthier future for all Angolans, especially in the regions most affected by water scarcity. As noted by Cajazeiras (2020), increased dependence on groundwater during droughts increases the possibility of water scarcity, as the storage capacities of surface water resources tend to decrease considerably.

In the current context, it is crucial that Angola strengthens elements that guarantee responsible water consumption, promoting practices that optimise and rationalise its use, especially in face of population growth and increasing demands on water resources. Raising awareness and implementing effective actions for the preservation and sustainable use of water are fundamental to ensuring a safer and healthier future for all Angolans, especially in the regions most impacted by water scarcity. It is important to consider that during periods of drought, dependence on groundwater increases, which raises the risk of water scarcity, since the storage capacities of surface water resources tend to decrease considerably, as observed by Cajazeiras (2020).

2.2.1. Effluent treatment

Effluent treatments are essential procedures for eliminating contaminants and pollutants from effluents produced by human activities before they are discharged into the environment. These processes aim to safeguard water quality and reduce the adverse impacts water pollution on aquatic ecosystems and public health. Advanced treatment technologies such as filtration methods and processes Biological methods are widely used to ensure that effluents comply with established environmental standards. This study proposes to carry out a literature review on physical-chemical and biological methods of effluent treatment, highlighting their general characteristics and the results found in the literature, as mentioned by Azevedo (2020). Effluent treatment plays a fundamental role in protecting the environment and promoting public health. This process can be divided into different stages: preliminary, primary, secondary and tertiary.

In the preliminary stage, solids are separated from the effluent, usually by means of screening and sieving, and fine or fibrous solids are removed. Primary treatment includes several stages, such as neutralising the effluent with chemicals and flocculation, which separates the liquid elements from the solids (Pellenz, 2018).

Secondary treatment is responsible for the additional removal of organic matter, followed by further decantation to separate the sludge and clarify the water. The final stage is sludge treatment, where the organic matter is transformed into more stable compounds such as water and carbon dioxide (Carvalcanti, 2020).

It is important to emphasise that there are different approaches to effluent treatment, such as the adsorption process, which involves the interaction of molecules from a liquid or gaseous medium on the surface of a solid. In addition, facultative systems, such as treatment ponds, are used in conjunction with other units to increase the efficiency of organic matter removal. These treatment stages and methods are essential for guaranteeing water quality and reducing the negative impacts of water pollution on aquatic ecosystems and public health, as emphasised by Oliveira (2020).

To ensure compliance with the rules and regulations relating to effluent treatment, many companies choose to outsource this process to specialised partners. This is especially common in areas without a sewage network, due to the low cost of implementation and the high efficiency of domestic effluent treatment (Cherubini, 2021).

This approach provides cost savings, legal compliance and allows companies to focus on their core activities, leaving environmental management to specialists. Processes such as aerobic digestion are favoured by the presence of dissolved oxygen, promoting the oxidation of organic compounds and resulting in the formation of by-products such as sludge, carbon dioxide and water (Carvalcanti, 2020).

Effluent treatment therefore plays a crucial role in environmental preservation and the protection of public health. There are different types and techniques of treatment, categorised as preliminary, primary, secondary and tertiary, the choice being determined by the pollution load and presence of specific contaminants. Outsourcing this process has become a popular option among companies, offering advantages such as cost reduction and compliance with legislation, as pointed out by Azevedo (2020).

2.2.2. Water scarcity and drought

Water scarcity and drought represent growing challenges in many regions of the world, as a result of various factors such as climate change, population growth and inadequate water management practices. These phenomena can cause severe impacts on the availability of water for human, agricultural and industrial use, as well as affecting the balance of aquatic ecosystems (Sassoma, 2013). Faced with these challenges, it is extremely important to implement conservation measures, adopt efficient water management practices and invest in adaptation technologies to mitigate the effects of water scarcity and drought. In rural areas, for example, where water infrastructure is limited and populations are exposed to recurrent droughts, it is common to find dams that dry up during periods of drought and fountains (wells) that, more often than not, provide brackish water, as observed by Cajazeiras (2020).

Based on the articles provided, it is clear that water scarcity and drought are critical issues not only in Angola, but also in other regions of the world. Water is an indispensable resource for life and sustainable development, and it is worrying that only 3 per cent of the available water is fresh and suitable for human consumption. Demand for water continues to grow, while pollution of freshwater sources also increases, exacerbating the problem of scarcity in many countries (Cajazeiras, 2020). In Angola, despite being considered a country with an abundance of water, the effective implementation of legislation and practices for the preservation of this resource faces significant challenges. The lack of progress on these issues makes integrated and effective water resource management difficult, especially in the face of climatic factors that can overload the water system. Integrated analysis on a regional scale, such as that carried out by Cajazeiras (2020), using socio-economic, infrastructure and meteorological data, is fundamental to establishing the intensity of the risk of water scarcity and aiding public authority decisions.

In this context, water supply and demand play a crucial role in water resource management, enabling the identification of areas at risk of water scarcity. This information is essential to inform decision-making and public policies aimed at tackling the challenges arising from water scarcity and drought (Sassoma, 2013). The preservation of water is essential for biodiversity and the maintenance of life on Earth, requiring the adoption of practical measures in everyday life to ensure the sustainable use of this resource. Awareness and collective action are needed to tackle challenges such as water scarcity and drought, ensuring the availability of water for future generations and promoting environmental sustainability. Direct water vulnerability impacts the quality of life and well-being of human populations, highlighting the importance of preserving landscape to maintain biodiversity and water resources, as noted by Cajazeiras (2020). (See fig. 2.7).

Figure 2.7- Water scarcity.

Source: Own authorship.

Therefore, water scarcity and drought represent urgent challenges that demand the implementation of effective public policies, the raising of public awareness and the adoption of sustainable practices to guarantee the availability and quality of water for all. Studies such as the one carried out by Araújo (2021), which involve flow network modelling and analysis of water demand management scenarios in regions of scarcity, are crucial for guiding water resource management and operation strategies and tackling these challenges effectively. In Angola, water scarcity is a growing challenge, especially in the south of the country, where record droughts have drastically reduced the availability of water. vital resource. Environmental degradation, the effects of climate change and poor management of water resources make this problem even worse, especially in a region where water is crucial for the production of products such as N'Gola beer in Lubango. Water scarcity threatens the source of exceptional quality water used in the production of this beer, located in Tundavala, 2,200 metres above sea level, which could have negative impacts on the local economy (Araújo, 2021).

This water shortage is not exclusive to Angola. Studies by the National Water and Sanitation Agency (ANA) indicate that Brazil's main river basins could face reductions of up to 40 per cent in water availability by 2040. Regions such as the North, Northeast and part of the Centre-West of Brazil are particularly vulnerable, which could result in stretches of intermittent rivers and affect water supply, hydroelectric power generation and subsistence agriculture (Cajazeiras, 2020).

To tackle water scarcity, it is crucial to strengthen water management practices throughout the country, especially in the most vulnerable regions. In addition, it is necessary to promote awareness of the importance of water conservation and the need to reduce unnecessary consumption. International co-operation can also play a significant role in supporting developing countries, such as Angola, to tackle this crucial challenge (Araújo, 2021). Water scarcity is a constant reality for the inhabitants of Brazil's semi-arid region, impacting not only human supply, but also social and economic development (Araújo, 2021).

2.3. Challenges and problems related to water resource management

The challenges related to water resources management cover a wide range of issues, from the scarcity of drinking water to the contamination of water bodies and the impacts of climate change.To tackle these complex challenges,it is necessary to adop integrated approaches and collaborative approaches that take into account the needs of local communities, the protection of aquatic ecosystems and the implementation of effective water management policies. Efficient water resource management should always seek to provide for the multiple use of water, ensuring that water is used sustainably and equitably to meet the needs of human populations, promote socio-economic development and preserve natural ecosystems (Monteiro, 2022). (See fig. 2.8).

Figure 2.8 - Water management challenges and problems.

Source: Author

Angola, despite having an abundance of water resources, faces the challenge of a shortage of piped drinking water, especially in peripheral communities in cities and rural areas. Lack of access to drinking water in these areas can result in diseases such as cholera and represents an urgent concern that needs to be addressed. Globally, water scarcity is a serious problem, affecting around 663 million people worldwide. Nadina, L. H. (2020).

This water scarcity is especially critical in arid and semi-arid regions, such as the centre-north of the African continent. Although Angola has satisfactory water reserves in some areas, such as in the centre-west, shared with countries such as the Democratic Republic of Congo, the Central African Republic and Cameroon, there are still regions in the country facing water stress. Water scarcity can manifest itself both in physical lack of availability, such as lack of access due to inadequate infrastructure or the inability of the responsible institutions to guarantee a regular supply (Cajazeiras, 2020) To tackle water scarcity in Angola, it is crucial to strengthen water management practices, especially in the southern provinces and coastal areas. This efficient management is essential not only for achieving water and sanitation targets, but also for boosting the country's economic development. The lack of drinking water can negatively impact the growth of local businesses, such as the N'Gola brewery, which is an important employer and source of income for the municipality of Lubango·

In addition, the occurrence of floods, inundations and droughts is directly linked to various natural disasters recorded in the country, particularly in Luanda, where frequent floods cause human losses and significant damage to infrastructure. In short, the scarcity of drinking water is a real challenge in Angola, especially affecting peripheral communities and rural areas. To tackle this problem, it is necessary not only to strengthen water management, but also to promote sustainable economic development and guarantee adequate and regular access to drinking water for all (Cajazeiras, 2020).

2.3.1. Drinking water supply

The supply of drinking water is a crucial issue for public health and human well-being, especially in Angola, where the water supply infrastructure faces significant challenges. Although there is reasonable access to drinking water in urban areas, the situation in Luanda, the capital, is worrying, with the majority of the population lacking adequate access to this vital resource. In rural areas, access to drinking water is also a concern, with investments underway to improve this situation in several towns. Nadina, L. H. (2020). (See fig. 2.9).

Figure 2. 9 - Drinking water supply

Source: Own authorship

Despite the country's considerable water potential, only a small fraction of it is being utilised for socio-economic development. Effective water management is fundamental to guaranteeing universal access to drinking water and promoting sustainable development. Lack of access to drinking water and adequate sanitation contributes to infectious diseases and represents a significant challenge to the health and well-being of the population. Nadina, L. H. (2020). Continued investment in infrastructure and effective management of water resources are needed to meet this complex challenge. Strategies for allocating and expanding water systems are fundamental to guaranteeing a reliable supply of drinking water at an affordable price, while at the same time preserving the ecosystem that produces this vital resource. These measures are essential for achieving sustainable development goals and guaranteeing a healthy and prosperous future for all Angolans (Cajazeiras, 2020; Lubanzadio, 2020; Araújo, 2021).

2.3.2. Impacts of climate change on water availability

Climate change has a significant impact on the availability of water around the world, with consequences that affect both the quantity and quality of this vital resource. Extreme weather events such as prolonged droughts and heavy rainfall are becoming more frequent and severe due to global warming caused by greenhouse gas emissions. These changes in climate directly affect the amount water available, as they can alter rainfall patterns, leading to more severe periods of drought in some regions and flooding in others. In addition, climate change also affects water quality, increasing pollution and contamination due to chemical run-off from human activities. Frascareli, D. (2021).Faced with these challenges, it is crucial to adopt adaptation and mitigation strategies to guarantee water security. This includes implementing public policies aimed at the sustainable management of water resources, promoting water conservation and reducing greenhouse gas emissions to mitigate the effects of climate change. Azevedo, P. G. (2020).

In addition, it is important to recognise the cultural value of water and its role in carbon sequestration, contributing to climate change mitigation. This highlights the need for integrated approaches that consider not only the physical and environmental aspects of water management, but also the social and cultural aspects (Santos, 2024).

Thus, tackling the impacts of climate change on water availability requires a coordinated and comprehensive response, involving action on several fronts, from reducing greenhouse gas emissions to implementing adaptation measures and sustainable management of water resources (Costa, 2022).

1. Climate Change and the Hydrological Cycle: To understand how climate change affects water availability, it is essential to understand how the hydrological cycle is impacted by these changes. This cycle involves a series of processes, such as evaporation, cloud formation, precipitation and surface runoff. As global temperatures rise, changes are expected to occur in all of these components, affecting the amount and distribution of water available in different areas. This is crucial to understanding the context of the current paradigm shift in the socio-environmental approach (Costa, 2022).

2. Impacts of Climate Change on Water Availability: Climate change is having a number of impacts on water availability. In some regions, rainfall patterns are changing, resulting in longer droughts or intense rainfall. In addition, the melting of glaciers and polar ice caps is contributing to rising sea levels and salinisation of aquifers, making fresh water scarcer. This directly affects socio-economic vulnerability and access to basic infrastructure (Silva, 2022).

3. Consequence for drinking water supply and agriculture: The scarcity of drinking water is a growing concern as a result of climate change. With less water available, especially in densely populated urban areas, access to drinking water is expected to be limited. This also affects agriculture, leading to falls in food production and increases in prices. Dependence on surface water sources further exacerbates this situation (Silva, 2022).

4. Socio-economic and environmental impacts: The impacts of climate change on water availability have serious socio-economic and environmental consequences. Vulnerable communities are the most affected, facing conflict and migration due to water scarcity. In addition, aquatic ecosystems suffer from reduced water flow and pollution, threatening biodiversity and ecosystem services. This emphasises the urgent need for action (Pereira, 2022).

5. Adaptation and Mitigation Measures: Faced with these challenges, it is crucial to take measures to adapt to the impacts of climate change and mitigate its consequences. This includes investments in resilient infrastructure, sustainable agricultural practices and reducing greenhouse gas emissions. It is important to act in a coordinated manner to guarantee a sustainable future for the next generations (Araújo, 2021; Pereira, 2022).

2.3.3. Sustainable technologies for water conservation

Sustainable water conservation technologies are fundamental to responsible and efficient water resource management. They encompass a variety of innovative solutions, from efficient irrigation systems to recycling and treatment technologies. These solutions offer creative ways of conserving and reusing water, reducing waste and minimising the environmental impact of human activities (Gramkow, 2020, Nolasco, 2020). An important example of these technologies in agriculture are drip and micro-sprinkler irrigation systems, which deliver water directly to plant roots, reducing losses through evaporation and surface run-off. In addition, practices such as no-till farming and mulching help to improve water retention in the soil, reducing the need for irrigation (Gama, 2020). (See fig. 2.10).

Figure 2.10 - Water recycling and effluent treatment technologies

Source: Own authorship

In the urban context, water recycling and effluent treatment technologies are essential to meet the growing demand for drinking water. These systems allow water to be reused for non-potable activities such as irrigation and toilet flushing, contributing to the conservation of water resources (Silva P. S., 2021). Rainwater harvesting is another technology on the rise, offering a free and sustainable source of water for various purposes. These systems can be installed in homes and commercial buildings, reducing dependence on conventional water sources (Alves, 2020).

In the industrial sector, water reuse technologies such as reverse osmosis systems and advanced effluent treatment allow companies to reduce water consumption and minimise waste generation, contributing to the conservation of water resources (Carvalcanti, 2020). In addition, the use of monitoring technologies and intelligent management of water resources helps to optimise the use of water in different sectors, allowing for a more efficient allocation of resources and a faster response to extreme weather events (Freitas, 2021).

In this way, sustainable technologies for water conservation play a fundamental role in promoting the responsible management water resources, contributing to the preservation and sustainable use of water for future generations (Carvalcanti, 2020).

2.4. Policies and legislation related to water resource management

Policies and legislation aimed at managing water resources play a fundamental role in defining guidelines and standards for the use and preservation of water. They address a wide range of issues, from integrated river basin management to regulating water use in different sectors. Robust legislation and its effective implementation are crucial to ensuring the sustainability of water resources and equitable access to water for all communities (Reis, 2020).

A central point of these policies is to promote integrated river basin management, recognising the interconnectedness of the elements within a basin and coordinating the use and protection of water resources. This involves the participation of various actors in decision-making and the implementation conservation measures (Gomes, 2020).

In addition, policies on water resources address specific issues related to water use in different sectors, such as agriculture, industry and public supply. They aim to encourage sustainable practices and set limits on water use (Matos, 2020). The protection of aquatic ecosystems and the conservation of biodiversity are also important aspects of these policies, involving the creation of protected areas and measures to reduce pollution of aquatic habitats (Frascareli, 2021).

Another important point is to guarantee fair access to water, especially in rural and peri-urban areas, through supply programmes and the transparent distribution of water resources (Costa, 2022).

Legislation can also establish management instruments, such as river basin plans and water use licences, as well as monitoring and inspection mechanisms to ensure compliance with regulations (Monteiro, 2022). Thus, policies and legislation related to water resource management play a vital role in promoting sustainability and protecting these resources, helping to guarantee access to water for present and future generations Reis, B. M. (2020).

The chapter on water resource management emphasises the vital importance of effectively managing one of our planet's most precious resources. By exploring topics such as the distribution, demand, preservation and sustainable use of water, it becomes clear that this management is not only an environmental issue, but also a social and economic one (Matos, 2020).

Understanding the geographical distribution of water and the different demands for it helps us to target effective management and allocation strategies for this vital resource. From domestic to industrial use, each sector has its own needs and challenges, requiring efficient water use and conservation practices Holmer, S. A. (2020). The pressing need for the preservation and sustainable use of water is evident in the face of the finiteness of this resource and emerging challenges such as water scarcity and drought. Strategies such as effluent treatment, efficient water management and sustainable technologies are essential to meet these challenges and guarantee the availability of clean water for future generations Gramkow, C. (2020). However, we face a number of challenges, from the scarcity of drinking water to the impacts of climate change on water availability. Integrated and collaborative management is key to addressing these problems, along with robust policies and legislation that guarantee equitable access to water for all communities Gama, E. M. (2020). Thus, water resource management is a complex task that requires a holistic and collaborative approach. Only through joint efforts can we guarantee the availability and quality of water for present and future generations, thus promoting environmental, social and economic sustainability.

CHAPTER 3
INTEGRATING ENVIRONMENTAL EDUCATION INTO WATER RESOURCES MANAGEMENT

In Chapter 3, we explore the integration of environmental education into water resource management, recognising the crucial importance of community awareness and engagement in promoting sustainable water use practices. We highlight various aspects related to environmental education and its application in water resource management, as well as strategies for involving the community in this process. Environmental education plays a fundamental role in making communities aware of the importance of conserving and using water responsibly. It is no longer about salvationist approaches, but about learning and recognition, providing broad meaning for thousands of people (Pereira, 2022, p. 452).

By integrating Environmental Education into water resource management, we seek to promote a deeper understanding of the interactions between society and the aquatic environment. This includes not only transmitting knowledge about the importance of water, but also encouraging active community participation in decision-making and the implementation of conservation and sustainable use measures (Alves, 2020).

It is essential to develop educational strategies that are contextually relevant and culturally sensitive, taking into account the specificities of each region and community Silva, M. H. (2022). In addition, we must take advantage of various educational tools and approaches, such as school programmes, public awareness campaigns, community workshops and the use of digital technologies, to reach a wider and more diverse audience. By strengthening the integration of Environmental Education into water resource management, we are empowering communities to become agents of change in favour of the conservation and sustainable use of water. This collaborative and participatory approach is fundamental to tackling the challenges related to the availability and quality of water, guaranteeing a safer and more sustainable future for all.

3.1.The role of environmental education in raising awareness of the importance of water

Raising awareness of the importance of water is fundamental to promoting its sustainable management. In this context, Environmental Education plays a crucial role, providing knowledge and sensitising the population to the relevance of water to human life, ecosystems and the economy. This discipline is not limited to the classroom, but seeks broader and more relational perspectives, offering a plural understanding of the world (Pereira, 2022, p. 465).

Environmental Education aims to rethink socio-cultural and socio-environmental relations, influencing human attitudes and actions in relation to the environment. In the search for sustainable water management, this discipline is essential for identifying problems related to the use of this vital resource (Cajazeiras, 2020, p. 31).

In the field of educational policies, programmes such as PRONEA and PNMA define guidelines for the inclusion of Environmental Education in schools, covering all levels of education. In addition, initiatives such as the "Water, your future in our hands" project aim to raise awareness and mobilise the school community about the importance and responsibility of water use (Almuiña, 2020, p. 15).

In Angola, Environmental Education is a relevant topic, although the research results do not specifically address this issue. However, studies on the Angolan education system and its relationship with the political context can provide important insights for promoting awareness about the importance of water and adopting responsible attitudes towards the environment (Almuiña, 2020, p. 8).

In short, Environmental Education plays a fundamental role in raising awareness about the importance of water, influencing human culture and behaviour in relation to the environment. This topic is worked on across the board in schools and is essential for promoting sustainable management of this indispensable resource (Cachapa, 2020, p. 53).

3.2. Strategy to involve the community in the sustainable management of water resources

The strategy to involve the community in the sustainable management of water resources is fundamental to guaranteeing the active and effective participation of the various actors involved.

Silva, D. C. (2021). Based on the documents provided, some approaches and practices cited by the author can be considered to promote this community participation:

1. Community Water Management Model (MoGeCA):

- MoGeCA prioritises the involvement of local groups in the negotiation, construction and management of water points as the basis for promoting water sustainability.
- Institutions such as Community Water Councils and Municipal Energy and Water Companies play essential roles in the sustainable management of water systems.

2. Community Participation in Water Quality Monitoring:

- Forming a network of water agents to assess the quality of water resources throughout the region is an effective strategy for involving the community in water monitoring and management.
- The Volunteer Environmental Agents Training Programme in schools plays a crucial role in training individuals who are committed to environmental preservation and water resource management.

3. National Water Resources Policy (PNRH):

- The PNRH provides for decentralised and participatory management of water resources, involving public authorities, users and communities.
- The active participation of organised civil society is essential to ensure effective community influence in water resource management processes.

In this way, the strategy for involving the community in the sustainable management of water resources must include the creation of participatory structures, such as community councils and training programmes for volunteer environmental agents, as well as ensuring the decentralisation of management and the active participation of the various actors, including civil society and water users Rosa, A. M. (2020).

3.3. Environmental education as a tool for changing behaviour towards water

Changing behaviour towards water is crucial to ensuring its conservation and sustainable use, and Environmental Education plays a key role in this process. By providing knowledge and sensitising people to the importance of water for human life and ecosystems, Environmental Education promotes awareness that is essential for adopting more responsible practices. Through educational approaches, it is possible to prevent water waste by encouraging the conscious use of this resource in domestic, agricultural and industrial environments (Pereira, 2022, p. 463). In addition, according to Bastos (2021), environmental education promotes reflection on the activities that negatively impact water resources, highlighting need for behavioural changes to preserve water quality and availability. Evaluation is a basic principle of this approach, making it possible to identify the effects caused by the inappropriate use water and to analyse the social changes needed to promote more sustainable management of water resources. By enabling individuals to understand their relationship with water and their environment, Araújo at all (2022) believe that environmental education promotes a culture of respect and responsibility towards natural resources. This awareness can extend beyond the individual sphere, influencing organisations and communities to adopt more sustainable practices in relation to water use. Thus, Environmental Education emerges as a powerful tool for promoting significant changes in the way society perceives, values and uses water, contributing to building a more sustainable and equitable future for all.

3.4. Challenges and opportunities in integrating environmental education into water resources management

Integrating environmental education into water resource management faces a number of challenges, but also presents significant opportunities. This subchapter analyses the obstacles and barriers encountered in implementing Environmental Education programmes aimed at water conservation, as well as the possibilities for overcoming them through innovative approaches, strategic partnerships and effective public policies. By tackling these challenges proactively, it is possible to harness the full potential of Environmental Education in promoting the sustainability of water resources. Almuiña, H. S. (2020). Integrating environmental education into water resource management is emerging as a promising strategy for raising awareness and transforming attitudes towards water use. However, the complexity of this process is evidenced by the various challenges faced. One of them is the lack of knowledge and awareness about the importance of preserving water resources, which can make

difficult to adhere to sustainable practices. In addition, resistance to changing behaviour and a shortage of human and financial resources are significant obstacles to implementing large-scale environmental education programmes Alves, K. d. (2020).
Another challenge is the need for coordination and cooperation between different entities and sectors involved in water management, which can create difficulties in implementing effective initiatives. Furthermore, measuring the impact and effectiveness of environmental education programmes in changing behaviour towards water is a complex issue, which makes it difficult to assess the success of these interventions (Cachapa, 2020).
However, despite these challenges, there are opportunities to improve the integration of environmental education into water management. The use of technologies and social media can be an effective strategy for disseminating knowledge and messages about the importance of sustainable water management, reaching a wider and more diverse audience. These opportunities can help strengthen community awareness and engagement in promoting sustainable water use practices. Partnerships and collaborations between governments, non-governmental organisations, companies and communities to implement environmental education programmes (Cajazeiras, 2020).
Incorporating themes related to water and sustainable management into school programmes and curricula. Training professionals and community leaders to promote environmental education and behaviour change in relation to water (Guerra, A.2020). Monitoring and continuous evaluation of the impact of environmental education programmes on changing behaviour in relation to water.
Integrating environmental education into water resource management can be an effective strategy for promoting awareness and changing behaviour in relation to water use. However, it is necessary to overcome challenges such as lack of knowledge and awareness, resistance to behaviour change, lack of human and financial resources, lack of coordination and cooperation, and difficulties in measuring the impact environmental education programmes. Opportunities include the use of technologies and social media, partnerships and collaborations, incorporating water-related topics into school programmes, capacity building and training for professionals and community leaders, and continuous monitoring and evaluation of the impact of environmental education programmes (Cherubinil, 2021).
The chapter on Integrating environmental education into water resources management highlights crucial importance of community awareness and engagement in promoting sustainable water use practices. By exploring various aspects related to environmental education and its application in water resources management, this chapter reveals key strategies for involving the community in this process Pereira, V. A. (2022).Environmental education has emerged as a powerful tool for raising awareness about the importance of water and encouraging responsible use and conservation practices. By empowering individuals with knowledge and awareness, we can promote a significant change in behaviour towards water, which is essential to ensure its conservation and sustainable use (Azevedo, 2020). The strategy for involving the community in the sustainable

management of water resources highlights the importance of active participation, the development of educational programmes and the creation of partnerships between different actors. This inclusive and collaborative approach aims to promote more efficient and inclusive management of water resources, meeting present needs without compromising future generations. Almuiña, H. S. (2020).

Although we face challenges in integrating environmental education into water resources management, such as obstacles in implementing educational programmes and political barriers, we also find significant opportunities Pereira (2022). Innovative approaches, strategic partnerships and effective public policies can overcome these challenges and maximise the potential of Environmental Education in promoting the sustainability of water resources.

Thus, integrating environmental education into water resource management is essential to guarantee a sustainable future for present and future generations. By empowering the community and promoting a change in behaviour towards water, we can work together to conserve this vital resource and guarantee its availability for human and environmental needs.

CHAPTER 4

EDUCATIONAL EXPERIENCES AND PRACTICES ENVIRONMENTAL AND WATER RESOURCES

In Chapter 4, a detailed analysis will be made of various experiences and practices in Environmental Education related to Water Resources. This examination will cover case studies, initiatives and pedagogical approaches that focus on awareness-raising and sustainable water management. Environmental education plays a crucial role in this context, as water is essential for life and ecosystems, requiring not only practical action, but also a change in mentality and behaviour on the part of society.

Initiatives will be examined that emphasise the importance of raising awareness about water, not just as a natural resource, but as a vital element for environmental sustainability and human well-being. These educational practices aim not only to inform, but also to inspire concrete actions for the conservation and responsible use of water, involving as many individuals as communities Matos, R. P. (2020). In addition, case studies will be presented that demonstrate how Environmental Education can be applied effectively in different contexts, from local communities to school and institutional settings. These examples will provide relevant data on best practices, challenges faced and lessons learnt in promoting awareness and sustainable management of water resources. Throughout the chapter, the role of environmental education as a catalyst for positive change will be emphasised, not only in the individual sphere, but also in the collective sphere. By highlighting successful experiences and innovative practices, the aim is to inspire and empower readers to become actively involved in protecting and preserving this vital resource for life on the planet (Pereira, 2022). In this context, Chapter 4 sets out to analyse case studies that highlight water-related environmental education programmes. These studies exemplify successful initiatives that not only inform, but also promote active public participation in the protection of water resources. By examining these cases, it will be possible to understand how these programmes engage individuals and communities, encouraging the adoption of sustainable practices and raising awareness about the importance of water for ecological balance and human well-being.In addition, various pedagogical approaches to water management education will be explored. This will include analysing methods, tools and developing educational curricula that focus on water appreciation and conservation practices. These approaches aim not only to impart knowledge, but also to stimulate a change of mentality in relation water use, promoting the adoption of more responsible and sustainable behaviours.

By sharing these experiences and practices, the aim is to inspire and empower individuals and communities to play an active role in the preservation and sustainable use of this vital resource for our planet. Low reservoir levels present a unique opportunity for ecological maintenance and the quality of water abstracted, highlighting the importance of educational initiatives that promote awareness and action in favour of protecting water resources. (Araújo, 2021, p. 22)

4.1. Case studies on water-related environmental education programmes

Case studies dealing with water-related environmental education programmes offer a detailed analysis of specific initiatives in this field. By examining these cases, it is possible to gain an in-depth understanding of how such programmes are designed, implemented and evaluated in real-life scenarios. They provide a comprehensive overview of the challenges faced, the successes achieved, the failures encountered and the lessons learned throughout the implementation of these programmes. These studies are essential for a more complete understanding of how environmental education is applied in practice and what its impact is on environmental awareness, behaviour change and the sustainable management of water resources (Reis, 2020).
Within the framework of these case studies, crucial issues are explored, such as the inclusion of different demographic groups, including women, in water management initiatives. For example, a study might analyse the barriers and opportunities women face when participating in councils or associations involved in water management. This analysis can reveal challenges related to existing power relations between management members and provide relevant information on how to promote more equitable and effective participation by all community members in water decision-making (Silva P. S., 2021).
These case studies not only highlight the importance of environmental education in promoting sustainable water management, but also identify areas in need of improvement and enhancement. They serve as a source of inspiration and guidance for future environmental education initiatives and policies, helping to shape more effective and inclusive strategies for dealing with contemporary environmental challenges, especially with regard to the conservation and responsible use of water resources. (Rosa, 2020, p. 117).

4.2. Examples of initiatives that promote public participation in the protection of water resources

These examples highlight various initiatives that seek to actively involve the community in the protection and sustainable management of resources programmes. They can include volunteer programmes to clean up rivers and beaches, awareness campaigns on responsible water use, citizen-led water quality monitoring projects and partnerships between local governments, non-governmental organisations and companies to conserve watersheds. These initiatives demonstrate how public participation can be effective in preserving water resources, promoting awareness, engagement and shared responsibility for protecting the aquatic environment (Pinho, 2021).

4.3. Pedagogical approaches to water management education

The approaches mentioned refer to the educational methodologies and strategies used to teach and promote awareness about the proper and sustainable management of water resources. This can include implementing environmental education programmes in schools, universities and communities, as well as holding workshops, lectures and practical activities that address water-related issues. In addition, the development of specific educational materials centred on the responsible management of water resources is proposed Monteiro, M. J. (2022).
To improve the use of groundwater resource indicators in rural areas, measures such as strengthening the groundwater database, increasing institutional liaison and defining technical criteria when choosing communities are suggested. These measures aim to empower local communities to better understand the situation of water resources in their areas and to adopt more sustainable water management and conservation practices (Guerra, 2022).
These educational approaches aim to empower individuals and communities to understand the importance of water and to adopt practices for the conservation and sustainable use of this vital resource. By providing knowledge and promote awareness of the importance of water and best practices for its management, these educational initiatives can play a key role in fostering a more balanced and responsible relationship with this essential resource (Cajazeiras, 2020, p. 118).

4.3.1. Methods and tools used to teach about the importance of water and conservation practices

In the context of education about the importance of water and its conservation practices, a variety of methods and tools are employed to facilitate learning and promote a comprehensive understanding of these issues. Traditional educational methods, such as lectures and debates, offer opportunities for transmitting theoretical and conceptual knowledge about water, its cycles, uses and conservation challenges. (Silva P. S., 2021) In addition, practical activities such as laboratory experiments and field projects give students tangible, hands-on experience, increasing their understanding of the importance of water in the real world. In parallel, modern educational tools, including instructional videos, interactive apps, infographics and educational games, provide a more dynamic and engaging approach to learning, captivating students' interest and facilitating information retention. These combined methods and tools play a crucial role in educating about the importance of water and promoting conservation practices, empowering individuals to become active advocates for the sustainable management of water resources (Costa, 2022).
The integration of environmental education into sustainable water management can be achieved through the development of curricula, subjects and educational content centred on this theme. The Referential for Environmental Education for Sustainability, drawn up by the Directorate-General for Education in Portugal, is an example of a

document that can be used as a reference. for the creation of educational programmes that address sustainable water management (Rodríguez-Guerra, 2020). The framework covers topics such as the proper management of natural resources, including water, and has the Referencial de Educação Ambiental para Sustentabilidade (Environmental Education Framework for Sustainability) as its reference curriculum document. Environmental education is a compulsory cross-curricular area all levels and cycles of schooling, and aims to raise awareness of the importance of water and the need for its sustainable management (Pinho, 2021).

The teaching of environmental education can be integrated into different subjects, such as natural sciences, geography, history and social studies, and can be adapted to different levels of education. Environmental education can be used to raise awareness about the importance of water, the quality of water, the activities that lead to water being wasted, the path of water, the use of water in households and also about activities that prevent water being wasted (Almuiña, 2020).

Environmental education can be used to promote behaviour change in relation to water, taking into account the importance evaluation as one of the basic principles of environmental education. Evaluation can help identify the impacts and effects caused, identify the problem and analyse the social changes that have taken place.Nadina, L. H. (2020).

Environmental education can be used to promote awareness the importance of water and the need for its sustainable management, and can be integrated into different subjects and levels of education. The development of curricula, subjects and educational content focused on sustainable water management can help promote awareness of the importance of water and the need for its sustainable management, and can be an effective tool for promoting behaviour change in relation to water (Cachapa, 2020).

4.3.2. The importance of community participation in water education

Community participation is essential for education about the importance of water and its sustainable management. The community can play an active role in environmental education by participating in educational activities such as workshops, seminars, lectures and community events Sampaio, R. J. (2021).

Community participation can help promote awareness the importance of water and the need for its sustainable management, and can be an effective tool for promoting behaviour change in relation to water. The community can also be involved in water quality monitoring and assessment activities, which can help identify local challenges and opportunities for sustainable water management (Souza, 2020).

Community participation can be encouraged through the creation of educational programmes that involve the community, such as community environmental education programmes, environmental education programmes in schools and environmental education programmes for companies. Community participation can help ensure that education about the importance of water and its sustainable management is relevant and adapted to local needs (Holmer, 2020).

4.3.3. The importance of inter-institutional collaboration in water education

Inter-institutional collaboration is essential for education about the importance of water and its sustainable management. Collaboration between different institutions, such as governments, non-governmental organisations, companies and universities, can help ensure that water education is comprehensive, relevant and effective (Pereira, 2022). Inter-institutional collaboration can help ensure that water education is based on up-to-date scientific and technical knowledge, and that it is adapted to local needs. Inter-institutional collaboration can also help to ensure that water education is integrated into different sectors, such as agriculture, industry and energy, which can help to promote the sustainable management of water resources (Nolasco, 2020). Inter-institutional collaboration can be encouraged through the creation of collaborative networks, such as environmental education networks, research and development networks and action networks. Inter-institutional collaboration can help ensure that water education is a shared responsibility between different institutions, which can help promote the sustainable management of water resources (Nadiana, 2020).

4.3.4. The importance of evaluation and monitoring water education

Evaluation and monitoring are essential for education about the importance of water and its sustainable management. Evaluation can help identify the impacts and effects caused by water education, and can help identify opportunities and challenges for improving water education Reis, A. C. (2020). Monitoring can help ensure that water education is based on up-to-date and accurate data, and that it is adapted to local needs. Monitoring can also help ensure that water education is integrated into different sectors, such as agriculture, industry and energy, which can help promote the sustainable management of water resources Neiman, A. R. (2022).Evaluation and monitoring can be carried out using different methods, such as surveys, interviews, observations and data analyses. Evaluation and monitoring can help ensure that water education is effective and efficient, and that it is adapted to local needs.

4.3.5. The importance of innovation and creativity in water education

Innovation and creativity are essential for education about the importance of water and its sustainable management Holmer, S. A. (2020). Innovation and creativity can help ensure that water education is relevant, engaging and effective. Innovation and creativity can be encouraged by creating educational programmes that promote creativity and innovation, such as project-based environmental education programmes, game-based environmental education programmes and arts-based environmental education programmes. Innovation and creativity can help ensure that water education is an effective tool for promoting behaviour change in relation to water (Santos, 2024).

4.3.6. The importance of sustainability in water education

Sustainability is essential for education about the importance of water and its sustainable management. Water education should be based on sustainability principles, such as respect for ecosystems, social equity and economic efficiency (Alves, 2020).Water education should promote awareness of the importance of water for sustainability, and should promote the sustainable management of water resources. Water education should be integrated into different sectors, such as agriculture, industry and energy, which can help promote the sustainability of water resources. Water education should be a shared responsibility between different institutions, such as governments, non-governmental organisations, companies and universities, which can help promote the sustainability of water resources. Water education should be an effective tool to promote behaviour change in relation to water, and should be a shared responsibility between different actors, such as individuals, communities, governments and companies (Cajazeiras, 2020).

4.3.7. Development of curricula, subjects and educational content centred on sustainable water management

Developing curricula, subjects and educational content centred on sustainable water management involves creating teaching materials that specifically address the importance of water and conservation practices. This includes the careful selection relevant topics, such as the conservation of water resources, the water cycle, the responsible use of water and threats water quality. In addition, it is essential to develop practical activities and projects that allow students to understand these concepts in a tangible way. The aim is not only to inform about the importance of water, but also to inspire changes in behaviour and action towards the preservation of water resources (Araújo, 2021).
The integration of environmental education into sustainable water management can be achieved through the development of curricula, subjects and educational content centred on this theme. The referential for environmental education for sustainability, drawn up by the Directorate-General for Education in Portugal, is an example of a document that can be used as a reference for creating educational programmes that address sustainable water management. The referential covers topics such as the proper management of natural resources, including water, and has the Referential for Environmental Education for Sustainability as its reference curriculum document. Environmental education is a compulsory cross-curricular area at all levels and cycles of schooling, and has with the aim of raising awareness of the importance of water and the need for its sustainable management (Cherubini, 2021). The teaching of environmental education can be integrated into different subjects, such as natural sciences, geography, history and social studies, and can be adapted to different

levels of education. Environmental education can be used to raise awareness about the importance of water, the quality of water, the activities that lead to water being wasted, the path of water, the use of water in households and also about activities that prevent water being wasted. Environmental education can be used to promote behaviour change in relation to water, taking into account the importance of evaluation as one of the basic principles of environmental education. Evaluation can help to identify the impacts and effects caused, identify the problem and analyse the social changes that have occurred (Cajazeiras, 2020).

The integration of environmental education into water resource management, as explored in this chapter, reveals the fundamental role that community awareness and engagement play in promoting sustainable water use practices Guerra, F. S. (2022). By analysing various environmental education experiences and practices related to water resources, it was possible to highlight case studies, initiatives and pedagogical approaches aimed at raising awareness and sustainable water management.

5. METHODOLOGICAL DESIGN

5.1. Introduction

Qualitative research of a descriptive nature will be conducted, using an action research approach and with a cross-sectional duration. This section will present the methodological procedures adopted, emphasising the main elements of the study.

5.2. Variables

Based on the conceptual definitions presented in the theoretical framework, the operational variables to be investigated will be defined. The variables to be studied are:

- ✓ Levels of environmental awareness.
- ✓ Practices for the preservation/conservation of water resources.
- ✓ Impact of human activities on water quality.
- ✓ Existence of environmental education programmes.

5.3. Population and Sample

The sample for this research will be diverse, with the aim of capturing a wide range of perspectives on the environmental issues under study. We will include one hundred and fifty (150) students from three (3) general education schools in the town of Cuquema, which represents a significant sample of the region's student population. In addition, fifteen (15) employees of the local administration and twenty-five (25) employees of the aggregates mining company MSTR will also be part of the sample, providing valuable data on the perspectives of the community and the private sector. It is essential to emphasise that, in order to ensure equitable representation, the questionnaire will be administered to seventy-five (75) students from the three selected schools, as well as eight (8) employees of the local administration and ten (10) employees of the MSTR company in the same locality. This stratified sampling approach will enable a more detailed analysis of the responses and a more accurate understanding of environmental perceptions and practices within these specific groups.

5.4. Measuring instruments and techniques

To assess environmental awareness and water conservation practices, we will use questionnaires with closed questions on a Likert scale. These questionnaires will be distributed among several participants in the study area, covering the target population of this research on environmental education and water resource management, as detailed in the annex provided. We will ensure the confidentiality of the data and respect for ethical aspects during the application of the questionnaire, guaranteeing the integrity and privacy of the answers collected.

In addition to collecting data through the questionnaire, a talk will be given to the students of the three selected schools. The aim of this talk will be to promote environmental awareness and discuss the importance of preserving water resources. This initiative will also encourage teachers to integrate environmental issues into their academic activities, aiming for a more holistic and comprehensive approach to environmental issues in education.

5.5. Procedures

To ensure effective data collection, specific procedures will be adopted according to the characteristics of the participants and the nature of the research. The questionnaire will be the main instrument used, applied The questionnaire will be administered face-to-face to the students of the three selected schools and will then be entered by the author into the Google Forms platform. For the employees of the administration and the MSTR company, the questionnaire will be administered in person, using the Google Forms platform.

The semi-structured interviews will be conducted face-to-face with the school students, allowing for a more direct and in-depth approach to the issues raised in the questionnaire. On the other hand, for the employees of the Administration and the MSTR company, the interviews will be conducted in a non-face-to-face manner, taking into account the specificities of the participants and using their institutional representatives as intermediaries.

The approach to the participants will be adapted according to their specific characteristics and environments, ensuring the effectiveness and quality of the data collected. Participants will be identified through their institutional representatives, who are recognised as influential figures in the study locality.

The Google Forms electronic platform will be used to statistically analyse the data collected.

5.6. Hypotheses

The Cuquema community has a low level of environmental awareness about the importance of preserving water resources, which results in ineffective conservation practices and an overall negative perception of water quality, necessitating educational interventions and more robust public policies to promote sustainability and community participation in water resource management.

6. RESEARCH RESULTS

6.1. Result of the enquiry

In the first question, which addresses the Cuquema community's awareness of the preservation of water resources, the majority of responses indicate disagreement with the importance of preserving these resources, as shown in Graph 1, where approximately 43% of respondents say they do not recognise their importance. These results suggest an insufficient level of awareness in the community, pointing to the need for educational and environmental preservation initiatives (see Figure 6.11).

Figure 6.11- Awareness of the Cuquema community regarding the preservation of water resources

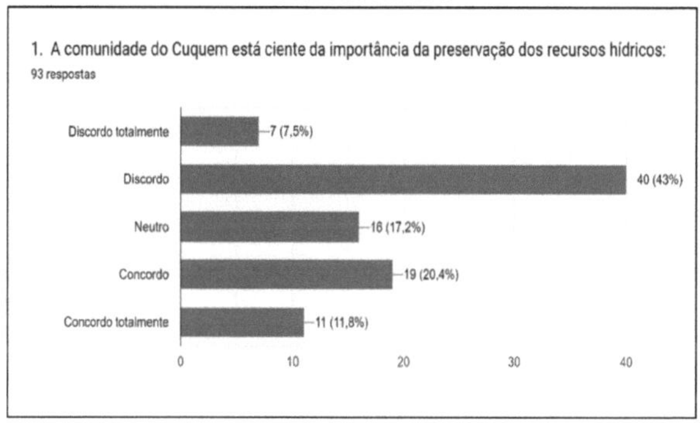

Source: Survey.

When asked whether the current practices of the Cuquema community contribute to the preservation of water resources, it was interesting to note that approximately 54.8 per cent of the participants disagreed with this statement. This indicates that a significant portion of the sample does not consider the current practices in preserving the region's water resources. This disagreement suggests the need for an analysis of existing practices or better communication about ongoing initiatives aimed at aligning community perceptions with preservation efforts. This finding can serve as a starting point for more in-depth investigations the community's concerns and opinions regarding preservation of water resources, providing valuable elements to guide future strategies and initiatives in this context (See Figure 6.12).

Figure 6.12- Current practices for preserving water resources in Cuquema

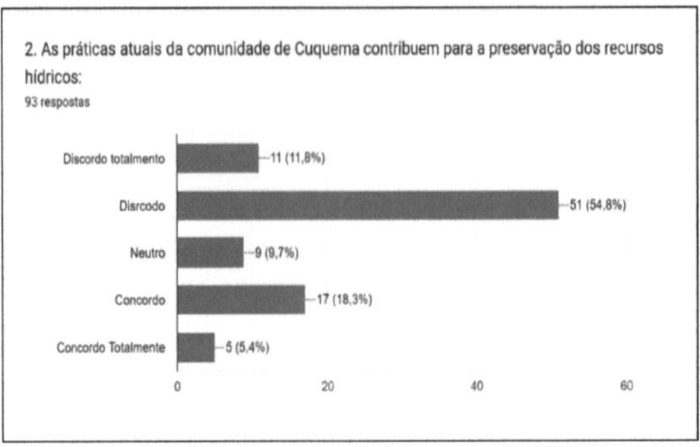

Source: Survey

When asked about the measures adopted by the Cuquema community to reduce water wastage, the results reveal a significant distribution of perceptions: 26.9 per cent consider these measures not effective, while 51.6 per cent see them as not very effective, 18.3 per cent think they are moderately effective, 3.2 per cent consider them very effective and 1.1 per cent think they are highly effective (See Figure 6.13). These responses highlight perceptible differences, as the majority perceive the measures adopted by the Cuquema community as not being very effective in terms of reduction in water wastage, as illustrated in Graph 3. This understanding suggests the need to re-evaluate existing strategies and look for new approaches that can be more effective in promoting awareness and implementing practices that lead to a significant reduction in water wastage.

Figure 6.13 - Effective measures to reduce water waste

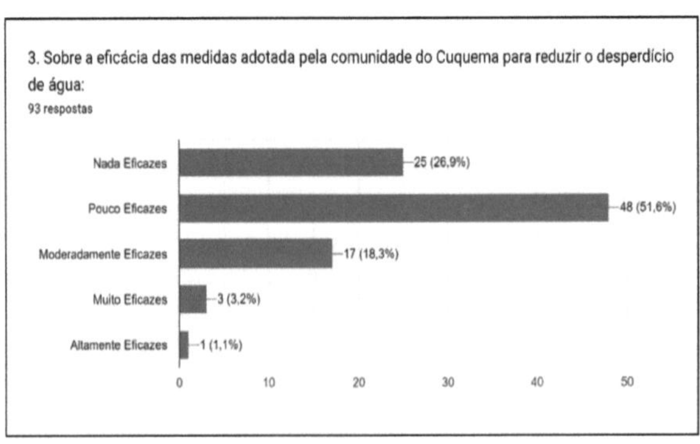

Source: Survey

According to the survey results, 40.9 per cent of participants disagreed, 23.7 per cent agreed and 21.5 per cent were neutral about the Cuquema community's concern water quality in their daily practices. These responses reflect a variety of perceptions within the sample, indicating divergent opinions on the community's level of concern about water quality.

The significant difference between the participants who disagree and those who agree, as well as those who are neutral, may indicate a lack of consensus or even dissatisfaction with existing practices related to water quality in the Cuquema community. On the other hand, the considerable number of neutral responses suggests a possible lack of clarity on this issue (See Figure 6.14).

Graph 6.14 - Concern about water quality

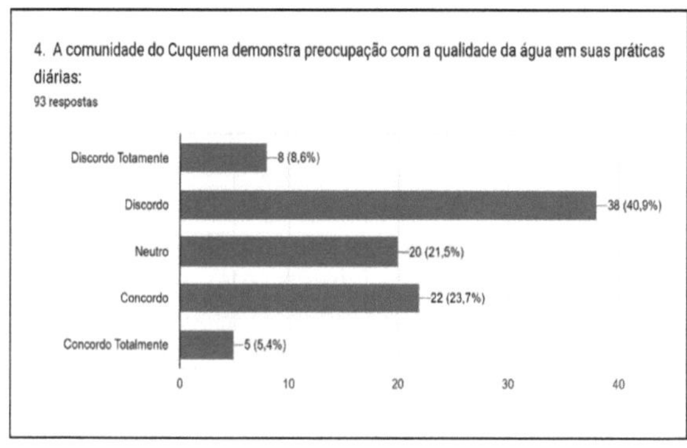

Source: Survey.

It is clear that approximately 54.8 per cent of the population questioned considers the Cuquema community to be insufficiently engaged, while 26.9 per cent perceive it to be not at all engaged in local water conservation initiatives, according to the results obtained in the field. These figures reflect a significant perception of the community's lack of engagement in water conservation initiatives (See Figure 6.15). This finding raises important questions about the underlying reasons for this low engagement, indicating the need for further investigation into the barriers, challenges and opportunities related to community participation in water conservation activities.

Figure 6.15 - Engagement in water conservation initiatives

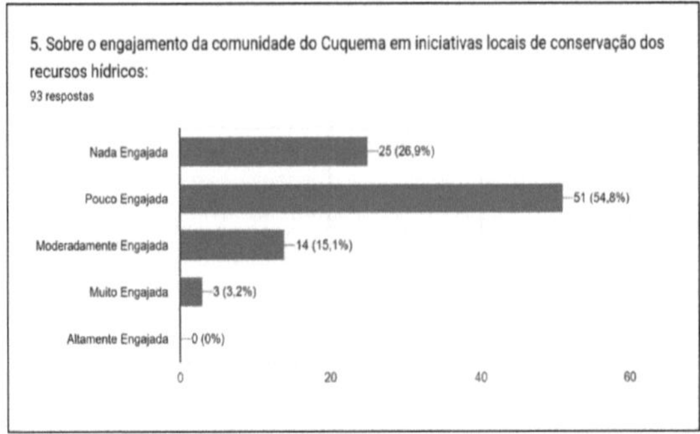

Source: Survey.

With regard to the frequency of practices or activities to preserve water resources by local authorities, the survey results show that 6.5 per cent of respondents believe that they never take place, 61.3 per cent say that they occur rarely, 29 per cent consider that they occur occasionally and only 3.2 per cent think that they occur frequently (See Figure 6.16).The survey results indicate that the majority of respondents perceive that local authorities rarely carry out practices or activities to preserve water resources. This suggests that local authorities are not as active in preserving water resources, which could be a point of attention for future policies and actions in this area. It is also important to note the percentage of those who perceive occasional frequency, indicating that there is still a significant portion of participants who recognise some activity, although not consistently.

Figure 6.16- Engagement in water conservation initiatives.

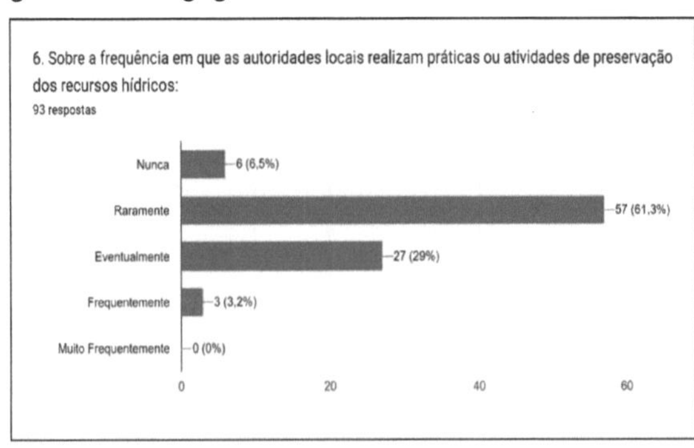

Source: Survey.

From the data collected, it is clear that the overwhelming majority of participants consider water to be extremely important for human survival and the health of the environment. The percentage of 90.3 per cent reflects a significant awareness of the importance of water, both for basic human needs and for preserving the environment (see figure 6.17).This result can serve as a basis for promoting environmental education campaigns and sustainability initiatives, with the aim of raising awareness about the responsible use of water and the importance of preserving aquatic ecosystems.

Figure 6.17- The degree of importance of water for human survival and the health of the environment.

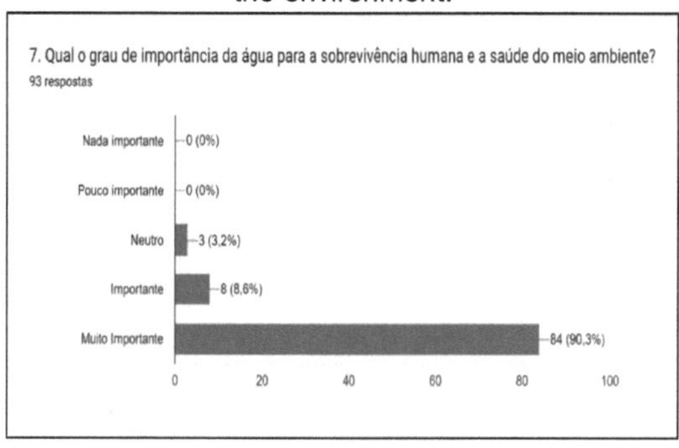

Source: Survey.

The results obtained on this question were very significant. The fact that 73.1 per cent of participants totally agreed that human activities have a significant impact on water pollution gives us a generalised perception of the influence of human actions on the quality of water resources (see Figure 6.18).
This high percentage of agreement reflects a substantial awareness of human responsibility in preserving water quality This perception is crucial for establishing policies aimed at mitigating pollution and protecting aquatic ecosystems.

Figure 6.18 - Impact of human activities on water pollution.

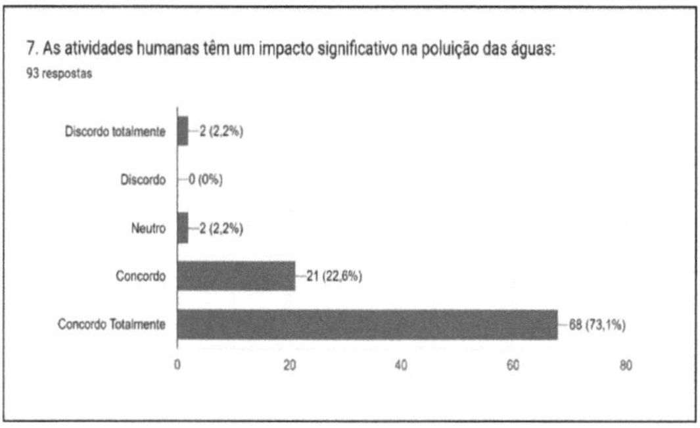

Source: Survey.

The scale used, as shown in the graph, provided a detailed overview of the participants' perceptions of their knowledge about the sources of water pollutionAnalysis of the results reveals that 36.6 per cent consider themselves to be very well informed on the subject, suggesting that the survey can make a significant contribution to increasing knowledge and awareness of the sources of water pollution.In addition, the fact that 31.2 per cent consider themselves moderately informed indicates that there is room to deepen their understanding. On the other hand, it's important to note that 12.9 per cent of respondents consider themselves to be poorly informed (see Figure 6.19).

Figure 6.19 - Sources of water pollution

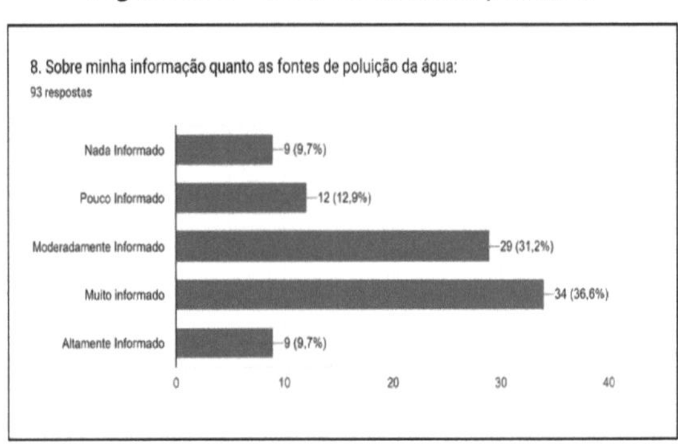

Source: Survey.

Based on the analysis of the data collected, it is clear that a considerable proportion of participants (48 per cent) fully agree that chemical use practices in agriculture have a direct impact on water pollution. This high percentage of agreement indicates significant knowledge of the relationship between agricultural practices and water quality, reinforcing the importance of addressing and mitigating these impacts. In addition, the 45.2 per cent of participants who agree with this statement also demonstrate a comprehensive recognition of the problem, albeit at a slightly lower level. The consistency of responses reflects a generalised awareness of the interconnection between agricultural practices and water pollution (see Figure 6.20).

Graph 6.20- Impact of agricultural chemical products on water pollution

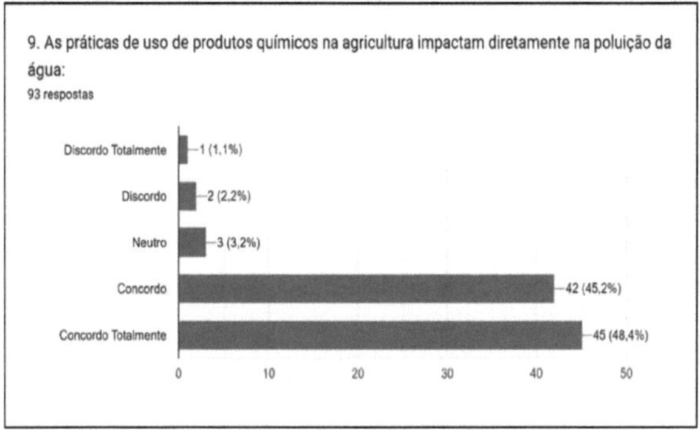

Source: Own authorship

The analysis showed that a significant proportion of participants (50.5 per cent) agreed with the statement, which indicates a generalised knowledge of the harmful effects of water pollution. This high percentage of agreement reveals a substantial awareness of the negative impacts of pollution on human health and the environment. In addition, the 44.1 per cent of participants who totally agreed with the statement show significant recognition of the harmful effects of water pollution (see Figure 6.21).These results provide a solid basis for advocating measures to protect and preserve water resources, as well as for promoting practices aimed at mitigating the harmful effects of water pollution. They also highlight the continued importance of educating and sensitising people about the environmental impacts of water pollution.

Figure 6.21- Effects of water pollution on human health and the environment

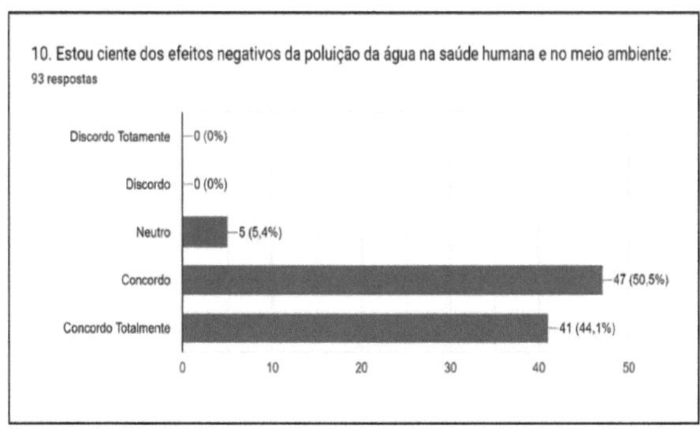

Source: Survey.

When asked about their level of agreement with the impact of inadequate solid waste disposal on water quality in the region, around 51.6 per cent totally agreed with the statement. This clearly shows that there is a significant relationship between the improper disposal of solid waste and water quality, demonstrating a keen awareness of the negative impacts of this problem.

In addition, 44.1 per cent agree with the statement, which indicates substantial recognition of the problem. The results obtained in this study allow us to work on more effective waste management practices and the implementation of policies aimed at preserving water quality (see Figure 6.22).

Figure 6.22- Effects of solid waste on water quality

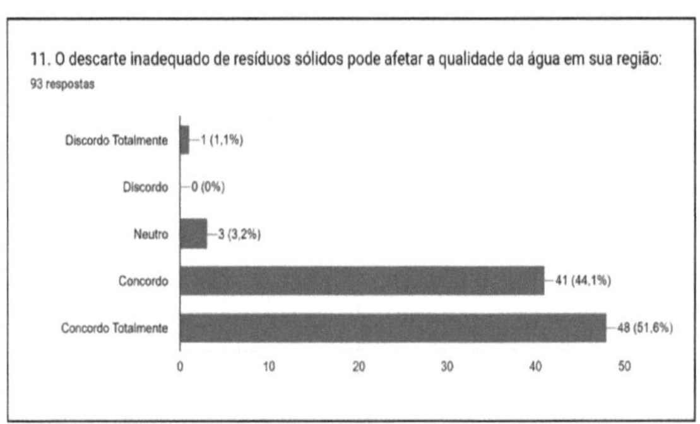

Source: Survey.

On the question relating to the level of information participants have about the sources of water pollution, the graph shows that a significant proportion of those interviewed consider themselves to be fairly well informed, which is positive as it indicates a sign of environmental awareness. However, it is important to note that there is still a considerable proportion who consider themselves only moderately informed (26.9 per cent) or not informed at all (12.9 per cent) (see Figure 6.23). This shows that there is room for improvement in publicising the sources of water pollution. Based on the results of this study, it would be interesting to explore which sources of information are most used by participants who consider themselves to be very informed, in order to understand which communication channels are being effective. In addition, it is crucial to investigate why those who consider themselves reasonably informed or not informed at all have this perception.

Figure 6.23 - Sources of water pollution

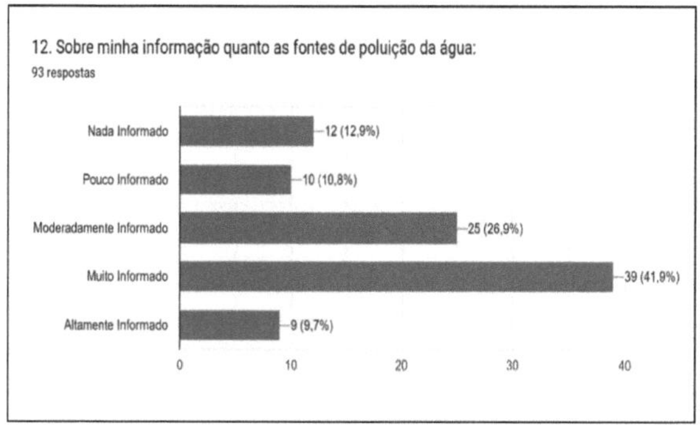

Source: Survey.

When asked which of the following anthropogenic activities has the greatest impact on water quality, the results show that 24.7 per cent of residents mentioned agricultural activities, 50.5 per cent pointed to solid waste dumping, and 18.3 per cent highlighted the inappropriate use of chemical products (see Figure 6.24). The fact that more than half of the participants consider the dumping of solid waste to be the anthropogenic activity with the greatest impact on water quality reveals a significant perception of the seriousness of this environmental problem. This leads us to reflect once on the urgent need to educate people about the effects of solid waste on water quality.

Figure 6. 24 - Anthropogenic activities with an impact on water quality

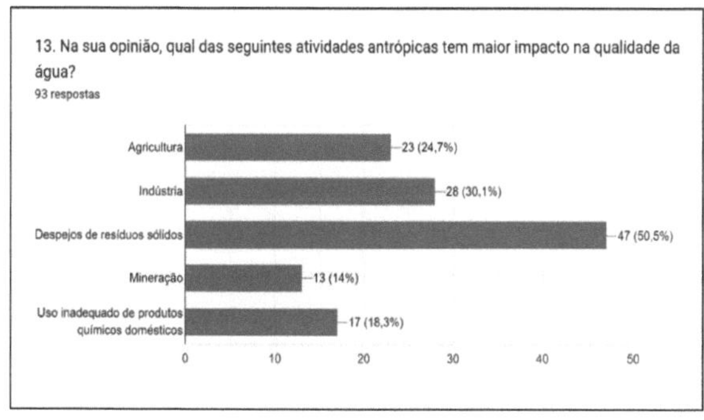

Source: Survey.

Concern about the impact of urban runoff on water quality is extremely relevant for the survey participants (54.8 per cent). Considering the urban and demographic growth observed Cuquema, it is clear how urban runoff has negatively influenced water quality. This phenomenon is responsible for introducing a variety of pollutants

such as oils, heavy metals, organic matter and chemical waste, which have a negative impact on rivers and aquifers (see Figure 6.25).
The data in Graph 15 reflects the interviewees' opinions on the impact of deforestation practices on water quality, totalling 93 responses. None of the interviewees totally disagreed that deforestation affects water quality. Only 3.2 per cent disagreed and 6.5 per cent were neutral about this impact. On the other hand, a significant majority of respondents (57 per cent) agree that deforestation has a significant impact on water quality, and 33.3 per cent totally agree with this statement. These results indicate that there is a general perception among those interviewed that deforestation is a harmful practice for water quality. With a total of 90.3 per cent (adding together those who agree and totally agree), there is a clear recognition of the direct relationship between deforestation and water quality degradation. This consensus reveals a high level of environmental awareness in the community about the negative impacts of deforestation. For public policies and environmental conservation initiatives, this data is fundamental, as it indicates a significant recognition of the environmental risks associated with deforestation, thus facilitating the implementation of environmental protection and restoration measures.

Figure 6.25 - Impact of urban runoff on water quality.

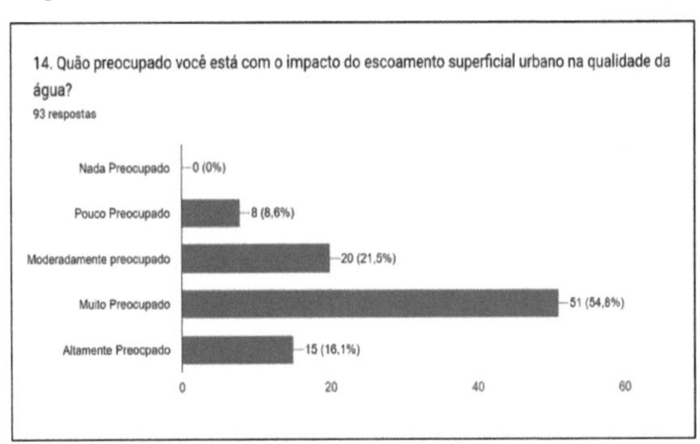

Source: survey.

More than half of the participants (57 per cent) agree that the practice of deforestation has a significant impact on water quality. In the locality studied, this practice is common and strongly influenced by the socio-economic conditions of the populations. Deforestation can result in the release of a variety of chemical substances and toxic materials into the environment, including water (see Figure 6.26).

Figure 6.26- Impact of deforestation on water quality.

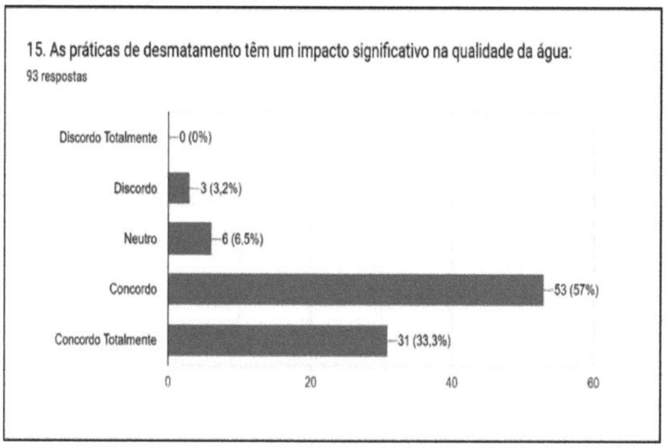

Source: Survey.

When asked about the activities they consider most harmful to water quality in the locality of Cuquema, the results reveal a variety of concerns among those interviewed. 31.2% believe that the inappropriate use of fertilisers and pesticides is the main threat, while 46.2% consider the inappropriate disposal of solid domestic waste to be the biggest problem. In addition, 23.7 cent point to poor basic sanitation infrastructure as a major concern, and 10.8 per cent each believe that oil and fuel spills and the improper disposal of hospital waste are significant (see Figure 6.27). This data indicates a varied awareness among residents of the factors that affect water quality in Cuquema. The greater concern about the improper disposal of solid domestic waste suggests that this problem is visible and frequent for the community. On the other hand, recognition of the impact of fertilisers and pesticides, as well as poor sanitation infrastructure, shows an understanding of less visible but equally harmful sources of water contamination. The lesser concern oil and fuel spills and hospital waste may reflect a lower incidence or visibility of these problems in the area, or a reduced perception of the risk they represent. These insights are valuable for guiding public policies and environmental education campaigns, focussing on areas that residents have already identified as problematic and raising awareness of other potential sources of contamination.

Figure 6.27- Harmful activities on water quality.

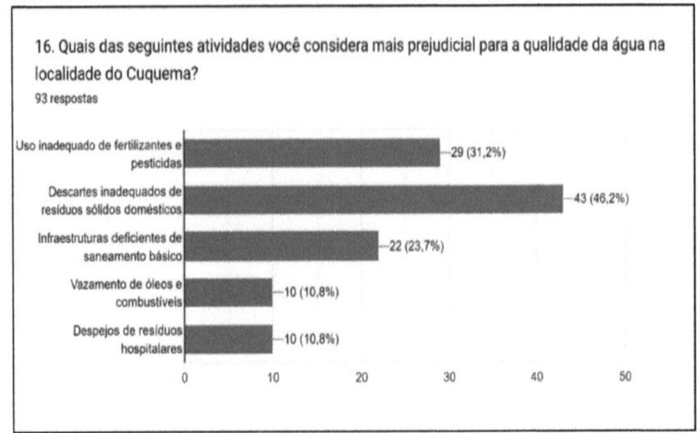

Source: Survey.

7. DISCUSSION

The survey results indicate that the majority of the Cuquema community is not aware of the importance of preserving water resources, with around 43 per cent of participants disagreeing with the relevance of this issue. This lack of awareness reflects a significant challenge in promoting sustainable practices. Environmental education emerges as a vital tool for reversing this scenario, as argued by Almuiña (2020) and Holmer (2020), who emphasise the historical and contemporary importance of environmental education in promoting sustainability.

As for current practices and the preservation of water resources, 54.8 per cent of those interviewed believe that the community's actions are not effective. This perception suggests the need to re-evaluate and improve existing practices, aligning them with sustainability principles discussed by Guerra (2022). The lack of effectiveness of current practices can be addressed by implementing more robust policies and environmental education initiatives that involve the community, as suggested by Costa (2022) and Neiman (2022).

With regard to perceptions of water quality, the survey reveals a significant divide in perceptions of water quality, with 40.9 per cent disagreeing with the concern for water quality, while 23.7 per cent agree and 21.5 per cent remain neutral. This diversity of opinions points to a lack of consensus on current practices and policies, which can be attributed to the insufficiency of effective educational programmes addressing water quality. Cajazeiras (2020) and Lubanzadio (2020) emphasise the need for public policies that guarantee the preservation of water resources and promote awareness of water quality.

The survey also sought to understand community engagement in conservation initiatives. The results showed a low perception of community engagement in conservation initiatives, with 54.8 per cent considering the community to be not very engaged and 26.9 per cent not engaged at all, suggests significant barriers to effective participation in conservation activities. This issue can be addressed through community involvement strategies and environmental education, as discussed by Pinho (2021) and Freitas (2023), who emphasise the importance of active community participation in water resource management perception that local authorities rarely carry out water conservation practices is a critical point that requires attention. This result emphasises the need for more active and transparent governance in the management of water resources. Sampaio (2021) and Nolasco (2020) argue that effective public policies are fundamental to guaranteeing the preservation of natural resources, emphasising the importance of government action in implementing conservation measures.

The data also shows that 90.3% of participants recognise the importance of water for human survival and the health of the environment, and 73.1% totally agree that human activities have a significant impact on water pollution. This awareness is a positive point and can serve as a basis for environmental education campaigns, as suggested by Azevedo (2020) and Araújo (2021). Promoting awareness of the interconnection between human activities and water quality is crucial for the development of sustainable practices.The majority of participants (48%) recognise the impact of agricultural chemicals on water pollution. This recognition emphasises the need for sustainable agricultural practices and the mitigation of the environmental impacts of agricultural activities. Frascareli (2021) and Freitas (2021) emphasise the importance of sustainable agricultural practices for conserving water resources.

The survey also highlights the perception that improper disposal of solid waste significantly affects water quality, with 51.6 per cent totally agreeing with this statement. This environmental problem can be addressed through effective waste management programmes and policies public programmes that promote sustainability, as discussed by Matos (2020) and Rodríguez-Guerra (2020).

The research reveals a series of challenges and opportunities in the management of water resources in the Cuquema river , Bié-Angola. The lack community awareness and engagement, combined with the ineffectiveness of current practices and insufficient government policies, underlines the need for an integrated approach that includes environmental education, effective governance and community participation. Authors such as Costa (2022), Neiman (2022) and Freitas (2023) provide a robust theoretical basis for the development of strategies that promote the sustainability and conservation of water resources.

Verification of the hypothesis in the discussion and results:

1. Discussion on environmental awareness:

- Low Awareness: The survey revealed that 43 per cent of participants disagreed with the importance of preserving water resources, indicating a significant lack of awareness in the community.
- Need for Environmental Education: Almuiña (2020) and Holmer (2020) argue that environmental education is crucial to promoting sustainability, supporting the need for educational campaigns to transform this perception.

2. Preservation practices:

- Ineffectiveness of current practices: 54.8 per cent of respondents believe that community preservation actions are not effective. Guerra (2022) suggests that sustainable practices need to be re-evaluated and improved.
- Policies and Education: Costa (2022) and Neiman (2022) advocate the implementation of robust policies and educational initiatives to improve preservation practices.

3. Perception of water quality:

- Diversity of opinions: The survey shows a lack of consensus on water quality, with opinions divided between disagreement, agreement and neutrality. This points to the insufficiency of effective educational programmes, as discussed by Cajazeiras (2020) and Lubanzadio (2020).
- The need for public policies: The lack of clarity about water quality highlights the need for public policies that guarantee the preservation of water resources and promote awareness.

4. Community engagement:

- Low engagement: With 54.8 per cent considering the community to be poorly engaged and 26.9 per cent not engaged at all, the survey suggests significant barriers to effective participation. Pinho (2021) and Freitas (2023) emphasise the importance of community involvement in conservation initiatives.

o Involvement strategies: The implementation of strategies to increase engagement, such as volunteer programmes and partnerships with local organisations, is essential.

5. Action by local authorities:

o Negative perception: The perception that local authorities rarely carry out preservation practices highlights the need for more active and transparent governance, as advocated by Sampaio (2021) and Nolasco (2020).

o Importance of government action: Effective public policies are fundamental to guaranteeing the preservation of natural resources.

6. Raising awareness about the importance of water:

o Positive Recognition: 90.3% of participants recognise the importance of water for human survival and the health of the environment. This data can be used as a basis for promoting environmental education campaigns, as suggested by Azevedo (2020) and Araújo (2021).

o Impact of human activities: The majority agree that human activities have a significant impact on water pollution (73.1 per cent total agreement), highlighting the need for pollution mitigation policies.

7. Sources of pollution and waste management:

o Recognising the Impacts: Most recognise the impact of agricultural chemicals and inadequate solid waste disposal on water pollution, indicating the need for sustainable agricultural practices and waste management programmes, as discussed by Frascareli (2021), Freitas (2021), Matos (2020) and Rodríguez-Guerra (2020).

The hypothesis that the Cuquema community has a low level of environmental awareness and ineffective conservation practices is verified by the survey results, which show a significant lack of awareness, negative perception of the effectiveness of conservation practices, diversity of opinions about water quality and low community engagement. Recommended interventions include the implementation of environmental education programmes, robust public policies and strategies to increase community engagement for sustainable water management.

8. GENERAL CONCLUSIONS

The research carried out into the environmental awareness of the population of Cuquema in relation to water resource management provided a comprehensive understanding of local perceptions and practices. By analysing the data collected, it was possible to achieve the overall objective of the study and identify crucial points for the formulation of future policies and actions.

The results show that there is a significant lack of awareness about the importance of preserving water resources. Around 43% of respondents disagreed with the importance of preservation, indicating the urgent need to implement environmental education initiatives to sensitise the community to the importance of water for human survival and the health of the environment. Educational campaigns are essential to change this perception. The majority of participants questioned the effectiveness of current preservation practices, with 54.8 per cent disagreeing with the effectiveness of community practices. This suggests the need to re-evaluate and improve conservation strategies, as well as to improve communication about ongoing actions. The introduction of new approaches and technologies can help to promote more effective water management practices and reduce water wastage.

The survey also highlighted a lack of consensus about the community's concern for water quality, with a diversity of opinions indicating dissatisfaction and a lack of clarity about current practices. Only 23.7% of participants agree that there is concern, underlining the need to strengthen awareness-raising initiatives and involve the community in practical conservation activities.

The low level of community engagement in conservation initiatives was highlighted by 54.8 per cent of respondents, pointing to the need to investigate the barriers preventing active community participation and developing strategies to increase engagement, such as volunteer programmes and partnerships with schools and local organisations.

The negative perception of the actions of local authorities, who rarely carry out preservation practices, highlights the need for more robust public policies and greater government intervention. Collaboration between authorities and the community is essential to implement effective preservation actions. On the other hand, the survey revealed a significant awareness of the importance of water, with 90.3% of participants recognising its relevance to human survival and the health of the environment. This positive data can serve as a basis for promoting environmental education campaigns that encourage the responsible use of water.

The majority of participants agree that human activities have a significant impact on water pollution, with 73.1 per cent in total agreement. This recognition is crucial for the development of pollution mitigation policies and the implementation of sustainable practices in agriculture and waste management. With regard to sources of pollution, although many participants consider themselves well-informed, there is still a significant proportion who do not have adequate knowledge. There is a need to intensify the dissemination of information about the sources of water pollution and the impacts of agricultural chemicals and improper disposal of solid waste.

The research also highlighted concerns about the impact of urban runoff and deforestation on water quality. This data reinforces the need to address the challenges of urbanisation and implement forest conservation policies as part of sustainable water management.

Thus, the results of the research indicate that although there is significant awareness of the importance of water and the impacts of human activities, there are gaps in conservation practices and community engagement. This data is fundamental for guiding future policies and actions to conserve water resources in Cuquema, promoting sustainable and participatory management that involves the entire community.

8.1. Recommendations

Based on the analyses and findings presented in this scientific paper, several recommendations can be made to promote sustainable water resource management and integrate environmental education in the Cuquema River Basin, Bié-Angola.

For the Provincial Government of Bié, it is suggested that robust policies and legislation be developed and implemented to promote the sustainable management of water resources in the Cuquema River Basin. These policies should cover effluent treatment, the prevention of water scarcity and the adoption of sustainable technologies in line with sustainable development goals. It is essential to establish an efficient monitoring and inspection system that regularly collects data on water quality and the impacts of human activities, ensuring compliance with the established policies.

Additionally, it is crucial to encourage community awareness and participation in water resource management. The government should organise awareness campaigns to inform the community about the impacts of human activities on water quality, using workshops, seminars and practical activities to engage the population. It is also recommended that local water management committees be set up, with representatives from different sectors of society, to ensure an inclusive and participatory approach to decision-making. In addition, it is recommended to offer incentives and subsidies for sustainable water use practices. Financial support should be made available for the implementation of technologies that save water and treat effluents, both for the community and for local industries. Such incentives can help foster a culture of sustainability and preservation of water resources in the region.

For the Ministry of the Environment in Bié, it is essential to promote environmental education and raise awareness of the importance of preserving water resources. Educational programmes focused on sustainable water management should be integrated into school curricula, and training programmes for educators and community leaders should be implemented to enable them to disseminate knowledge about water conservation.Creating partnerships and cooperation with non-governmental organisations, private companies and academic institutions is equally important. These partnerships can facilitate the implementation of water conservation projects by bringing in additional resources and technical expertise. In addition, it is necessary to encourage and fund research projects that explore new approaches and technologies for the sustainable management of water resources, promoting the dissemination of results to share good practices and lessons learnt.

For students studying biology, geography and natural sciences, it is recommended that environmental education be incorporated into the curriculum. Curricula and educational materials should be developed to address the importance of water and encourage the adoption of practical measures for its conservation, adapted to the local context of the Cuquema River Basin.

It is equally important to engage students in research and extension projects that explore new approaches and technologies for the sustainable management of water resources. Extension projects that involve the local community can promote awareness and the adoption of practices sustainable use of water. Finally, offering training and workshops that enable students to work in the field, collecting and analysing data on water quality and the impacts of human activities, is essential for developing practical skills and training qualified professionals in the field of environmental management.

9. BIBLIOGRAPHY

Almuiña, H. S. (2020). History of environmental education in Brazil and the world.

Federal University of Bahia, Institute of Biology.

Araújo, A. R., Guimaraes, B., Nascimento, M. de N. C. F., Melo do Vale, S. E. de J., & Santana Junior, V. C. (2022). Effectiveness of environmental education as a tool for behavioural change in Vila do Conde, Barcarena/PA. Brazilian Journal of Animal and Environmental Research, 5(2), 1747-1761. https://doi.org/10.34188/bjaerv5n2-024.

Alves, K. d. (2020). Gender and water. Catholic University of Minas Gerais - PUCMG.

Araújo, M. D. (2021). Vulnerability to water scarcity. Editora Amplla.

Azevedo, P. G. (2020). Physical and chemical processes for wastewater treatment: An integrative review. Federal Institute of Education, Science and Technology of Bahia, Salvador Campus.

Alberto, M. K. (2018). Environmental education in the pedagogical process in Angola. ROCA. Revista científico-educacional de la provincia Granma, 14(4), octubre-diciembre, 2018. ISSN: 2074-0735. RNPS.

Bastos, A. d. (2021). Environmental education and sustainability in degree programmes at the State University of Bahia, Campus VII. State University of Southwest Bahia.

Buza, R. G. C. (2013). Environmental education: Ideas, knowledge and practices reported by teachers in a country under reconstruction, Angola. Federal University of Pará.

Cachapa, A. F. (2020). The role of environmental education in the protection and valorisation of a natural resource: The case of the thermal waters of Montipa, Bibala-Angola.

Cajazeiras, C. C. (2020). Analysing vulnerability and risk to water scarcity in the semi-arid region: Case study Ibaretama/CE. Federal University of Ceará, Science Centre.

Carvalcanti, L. A. (2020). Physical and chemical processes for wastewater treatment: An integrative review. Federal Institute of Education, Science and Technology of Bahia.

Cherubini, K. G. (2021). Brief history of environmental education, its adoption in the Brazilian legal system and possible reflections of current political changes. State University of Southwest Bahia, Itapetinga Campus.

Costa, K. B. (2022). The applicability of environmental education in the protection and conservation of resources in Angola.

Frascareli, D. (2021). Spatial heterogeneity and influence of land use and occupation on the characteristics of surface sediment and interstitial water in the Itupararanga-SP reservoir.

Freitas, R. A. (2023). Environmental education and virtual environments from a critical perspective: The dynamics of cyberspace. Minas Gerais State University.

Freitas, W. B. (2021). Influence of land use and management on water infiltration: A

review. Federal University of Viçosa.
Gama, E. M. (2020). Analysis of alkalinity, chlorides, hardness, temperature and conductivity in water samples from the municipality of Almenara/MG. Federal Institute of Northern Minas Gerais, Almenara Campus.
Gomes, F. J. (2020). Analysis of alkalinity, chlorides, hardness, temperature and conductivity in water samples from the municipality of Almenara/MG. Federal Institute of Northern Minas Gerais, Almenara Campus.
Gramkow, C. (2020). Transformative investments for a sustainable development style.
Guarda, V. L. (2019). Gender and water. Federal University of Rio Grande do Sul - UFRGS.
Guerra, F. S. (2022). Landscape geoecology and applied environmental education: Foundations for environmental planning and management. Federal University of Ceará.
Holmer, S. A. (2020). History of environmental education in Brazil and worldwide.

Federal University of Bahia, Institute of Biology.

Lubanzadio, H. N. (2020). The national water plan in Angola: A diagnosis of the implementation of actions related to water supply. University of the International Integration of Afro-Brazilian Lusophony.
Matos, R. P. (2020). Analysis of alkalinity, chlorides, hardness, temperature and conductivity in water samples from the municipality of Almenara/MG. Federal Institute of Northern Minas Gerais, Almenara Campus.
Monteiro, M. J. (2022). Proposed actions for living with the scarcity of water resources: Approach to the municipality Santa Cruz do Capibaribe in Pernambuco. Federal University of Pernambuco, Agreste Academic Centre.

Nadina, L. H. (2020). The national water plan in Angola: A diagnosis of the implementation of actions related to water supply. University of the International Integration of Afro-Brazilian Lusophony.
Neiman, A. R. (2022). Principles and practices of environmental education.

Nolasco, G. M. (2020). Analysis of alkalinity, chlorides, hardness, temperature and conductivity in water samples from the municipality of Almenara/MG. Federal Institute of Northern Minas Gerais, Almenara Campus.
Oliveira, D. C. (2020). Physical and chemical processes for wastewater treatment: An integrative review. Federal Institute of Education, Science and Technology of Bahia.
Pereira, V. A. (2022). Environmental Otherness: Ontological contributions to the foundations of environmental education. Luis Amigó Catholic University.
Pinho, M. J. (2021). Environmental education and sustainability in undergraduate programmes at the State University of Bahia, Campus VII.
Reis, A. C. (2020). Analysis of alkalinity, chlorides, hardness, temperature and conductivity in water samples from the municipality of Almenara/MG. Federal Institute of Northern Minas Gerais, Almenara Campus.
Reis, B. M. (2020). Analysis of alkalinity, chlorides, hardness, temperature and conductivity in water samples from the municipality of Almenara/MG. Federal Institute of Northern Minas Gerais, Almenara Campus.

Rodríguez-Guerra, A. (2020). Social responsibility and environmental water management: A solution in Ecuador's dairy industry. Cordillera University Institute of Technology, Ecuador.
Rosa, A. M. (2020). Gender and water. Federal University of São João del-Rei - UFSJ. Sampaio, R. J. (2021). Brief history of environmental education, its adoption in the Brazilian legal order and possible reflections of current political changes. Federal University of Pernambuco.

Santos, A. P. (2024, April 22). A brief history of environmental education, concepts and their importance. RCMOS - Multidisciplinary Scientific Journal O Saber, 4(1). https://doi.org/10.51473/ed.al.v1i04.844
Silva, A. F. (2021). Environmental education and sustainability in undergraduate programmes at the State University of Bahia. State University of Southwest Bahia.

Silva, D. D. (2022). The implementation of public policies in the context of universalisation and tariff modicity in public drinking water supply and sewage services. Paulista State University.
Silva, E. V. (2020). Landscape geoecology and applied environmental education: Foundations for environmental planning and management. Federal University of Ceará.
Silva, M. H. (2022). The sustainable use of natural resources: An experience report from PIBID Bio. State University of Piauí.
Silva, P. S. (2021). Breve historia de la educación ambiental, su adopción en el ordenamiento jurídico brasileño y posibles reflexiones de los cambios políticos actuales. State University of Southwest Bahia.
Souza, F. R. (2020). Environmental education and sustainability: An emerging intervention at school.
Silva, D. C. (2021). Environmental management and sustainable development. Federal University of Maranhão.
Sassoma, I. T. L. (2013). Physico-chemical characterisation of water in the Catumbela River in Angola (Dissertation, Supervisor: I. F. de Souza).

GLOSSARY

Abbreviations or acronyms used in the text that are not widely used in Spanish will be recorded. In addition, terms that are unfamiliar, difficult to interpret or not commonly used in the context in which they appear will be included. Each of these terms must be duly defined or explained.

ANNEXES OR APPENDICES

Annexes are complementary documents that support or detail aspects cited in the body of the paper. Annexes should be numbered appropriately in this topic and correspond to their citation in the body of the paper.

Annexes 1- Research credentials

REPÚBLICA DE ANGOLA
GOVERNO DA PROVÍNCIA DO BIÉ
ADMINISTRAÇÃO MUNICIPAL DO CUITO
ADMINISTRAÇÃO COMUNAL DO CUITO

GABINETE DA ADMINISTRADORA ADJUNTA DA COMUNA SEDE DO CUITO

À

Coordenador do Centro Administrativo do Cuquema e Direção da Empresa MSTR

= Cuito =

NOTA Nº 19/GAA/JAC/04/2024

Melhores Cumprimentos!

Excelentissimos!

A Administração Comunal do Cuito, no cumprimento da **NOTA Nº 19/GAA/JAC/04/2024,** vem por esta via, encaminhar o **Sr. Abias Porfírio Cinco-Reis**, docente do Instituto Superior Politécnico do Bié, autorizado a realizar a sua investigação no âmbito da Educação Ambiental na Gestão dos Recursos Hídricos da bacia Hidrográfica do Cuquema, com o objetivo de avaliar o nível de conscientização ambiental da população do Cuquema em relação à gestão dos recursos hídricos, para a concretização dos seus objetivos, e aplicar os inquéritos no Centro Administrativo do Cuquema e na Empresa MSTR, prevista para o dia 23 de Abril de 2024, isto é, terça-feira.

Ciente de que o assunto merecerá atenção por vossa parte, nos despedimos reiterando votos de bom tralho.

GABINETE DA ADMINSTRADORA ADJUNTA COMUNAL DO CUITO, AOS 19 DE ABRIL DE 2024.

A Administradora Adjunta Comunal

Gonga

Centralidade, Q16, P19, Ap 0.2- Administração Comunal do Cuito- tel. 924640460

Appendices 2- Survey Questionnaire

ENVIRONMENT AND SUSTAINABLE DEVELOPMENT FIELD RESEARCH QUESTIONNAIRE

This questionnaire was drawn up as part of a master's dissertation in environmental management and auditing **at** the European University of **the** Atlantic entitled **Environmental education in the management of water resources in the Cuquema basin, Bié-Angola**, which is being developed by Abias Porfírio Cinco Reis, lecturer at the Higher Polytechnic Institute of Bié in Angola and supervised by Dr Leonardo Texeira, lecturer at the European University of the Atlantic.

Note: The data provided is absolutely confidential and anonymous and will be used exclusively for scientific research purposes. We ask you to be as rigorous as possible when filling them in.

Date the questionnaire was completed: / / Time: : Gender:Male ( ) Female ( ) Age

Marital status: Married ( ) Single ( ) Other ( )

Educational qualifications

Questions for the 1st specific objective: To analyse water resource preservation and conservation practices in the locality of Cuquema

5) **The Cuquem community is aware of importance of preserving water resources:**

( ) Strongly disagree ( ) Disagree (Neutral

( ) Agree

( ) Totally agree

6) **The current practices of the Cuquema community contribute to the preservation of water resources:**

( ) Strongly disagree ( ) Disagree ( ) Neutral

( ) Agree

( ) Totally agree

7) **On the effectiveness of the measures adopted by the Cuquema community to reduce water wastage:**

( ) Not at all effective ( ) Not very effective

( ) Moderately effective ( ) Very effective (Highly effective

8) **The Cuquema community shows concern for water quality in their daily practices:**

( ) Strongly disagree ( ) Disagree ( ) Neutral

( ) Agree

( ) Totally agree

9) **On the engagement of the Cuquema community in local water conservation initiatives:**

( ) Not at all engaged ( ) Not very engaged

( ) Moderately engaged ( ) Very engaged (Highly engaged

10) **On the frequency with which local authorities carry out practices or activities to preserve water resources:**

( ) Never

( ) Rarely

( ) Occasionally

( ) Often ( ) Very often

Questions for the 2nd specific objective: Identify the population's level of knowledge of the impacts of water pollution

11) **How important is water to human survival and the health of the environment?**

( ) Not important ( ) Not very important ( ) Neutral ( ) Important

( ) Very important

12) **Human activities have a significant impact on water pollution:**

( ) Strongly disagree ( ) Disagree ( ) Neutral

( ) Agree

( ) Totally agree

13) **About my information on the sources of water pollution:**

( ) Not at all informed ( ) Not very informed

( ) Moderately informed ( ) Very informed (Highly informed

14) **The use of chemical products in agriculture has a direct impact on water pollution:**

( ) Strongly disagree ( ) Disagree ( ) Neutral

( ) Agree

( ) Totally agree

15) **I am aware of the negative effects of water pollution on human health and the environment:**

( ) Strongly disagree ( ) Disagree ( ) Neutral

( ) Agree

( ) Totally agree

16) **Improper disposal of solid waste can affect water quality in your area:**
( ) Strongly disagree ( ) Disagree ( ) Neutral
( ) Agree

( ) Totally agree

17) **About my information on the sources of water pollution:**

( ) Not at all informed ( ) Not very informed ( ) Moderately informed
( ) Very informed

( ) Highly informed

Questions for the 3rd specific objective: Diagnose the main types of human activities with an impact on water quality

18) **In your opinion, which of the following anthropogenic activities has the greatest impact on water quality?**( )

Agriculture() Industry

( ) Solid waste dumping( ) Mining

( ) Inappropriate use of household chemicals

19) **How concerned are you about the impact of urban runoff on water quality?**
( ) Not worried at all ( )

Little concerned

( ) Moderately concerned ( ) Very concerned ( ) Highly concerned

20) **Deforestation practices have a significant impact on water quality:**

( ) Strongly disagree( )

Disagree ( ) Neutral ( ) Agree
( ) Totally agree

21) **Which of the following activities do you consider to be most damaging to water quality in Cuquema?**
( ) Inadequate use of fertilisers and pesticides ( ) Inadequate disposal of domestic solid waste ( ) Poor sanitation infrastructure
( ) Oil and fuel leaks( ) Dumping of hospital waste

We would like to thank you in advance for your co-operation and availability, and hope that this study will contribute to greater scientific knowledge on this subject.

Annexes 3- Deforestation in the Cuquema River Basin Region

Printed by Books on Demand GmbH, Norderstedt / Germany